CONTENTS

During 1979 supporting informations will be published on the following issues under the title "Siting Procedures - some national Cases":

- Social and Economic Impacts on Local Communities of the Siting of Major Energy Facilities: The Case of Nuclear Power Plants.
- The Role of Environmental Groups in the Siting of Major Energy Facilities: The U.S.A. Experience.
- Developments in Siting Procedures and Policies in the Federal Republic of Germany.
- Developments in the Procedures for Siting Thermal Power Plants in France.
- The Environmental Impact Statement. An Assessment of U.S.A. Experience.

PREFACE

Major energy facilities are amongst the most polluting industrial installations. Technologies are available, at some cost, to reduce adverse impacts from such installations. In addition, the socio-economic impacts have to be understood. This OECD report addresses such problems and refines the knowledge in the area of siting as presented in an earlier report, "Energy Production and the Environment", OECD, 1977.

The Report is primarily addressed to those who formulate national environmental or energy policies. It concentrates on siting policies and procedures and the division of responsibility between national, regional and local authorities. It also gives comprehensive information on socio-economic impacts resulting from the siting of major energy facilities.

To a certain extent, the study uses electricity generating plants as an example for describing the different phenomena. As the focus of the report is not primarily on physical impacts, the examples are valid also for such energy facilities as oil refineries, liquefied natural gas plants, oil terminals and coal synthetic fuel plants. Some data from these facilities are given. The report, however, does not examine problems of power transmission lines and pipelines as there is much ongoing work in these areas in other international bodies.

The study itself, and the conclusions it has been possible to draw from it, have been adopted as guidelines for Member countries for the siting of major energy facilities by the Environment Committee in April 1978.

The Secretariat of the OECD Nuclear Energy Agency and the OECD Combined Energy Staff have taken an active part in the preparation of relevant parts of this report.

SUMMARY AND GUIDELINES FOR THE SITING OF MAJOR ENERGY FACILITIES

STATEMENT OF THE PROBLEM

Major Energy Facilities (MEFs) are among the most polluting industrial installations. Technologies are available - at some cost - to reduce these emissions. Apart from the physical impacts, substantial socio-economic impacts have to be sustained by the local population. Most of these are caused by the influx of a large construction work force. Shortages of housing, schools, social services and communal services are the most common disamenities. Associated with these are many and important economic advantages: employment opportunities, increase in local business, and increased income from local taxation.

Extent of the problem for OECD Countries

On the basis of the OECD "World Energy Outlook" projections, the needs of OECD regions for new MEFs to meet energy demand by 1985 are presented.

The opposition to the siting of MEFs in OECD countries may be attributed to the following:

i) The high rates of industrialisation and the high density of population (national and regional) ;

ii) The necessity for most MEFs to be sited on coastal or riverine zones which, in many instances, are heavily industrialised and populated;

iii) The very large size of individual MEFs;

iv) The effects of their impacts (social, economic, environmental) being felt not only by the local community, but by the many surrounding communities as well;

v) The objection of a community to dramatic changes in the character of their social and cultural environment;

vi) The concern of citizens, and government, with the protection of the environment and the improvement in the quality of life; and,

vii) The demands of citizens to participate in decisions which are likely to affect their lives.

Nuclear power and the wider concerns it creates - radioactive waste management, nuclear proliferation, sabotage, etc. - has helped in mobilising the public in defence of environmental causes.

In recent years the opposition has been institutionalised in major national or international environment protection groups, and politicians and political parties have taken sides on the issues. Lastly, the oil price crisis obliged governments to take many of the responsibilities on energy policies which were previously left to the private sector.

The objective of this Report is to identify the siting problems and to provide information to Member governments on regulatory procedures and policies which may help in reducing the adverse environmental and social impacts of MEFs while allowing for the timely implementation of national and international energy policies.

MITIGATING THE ADVERSE IMPACTS OF MEFs

Part I. The Technical Approach

This section examines the technical options that are available - or are being developed - to the developers for reducing air and water emissions from electricity generating power plants.

With respect to the flexibility in siting provided by the use of emission controls the report makes the following points :

i) In many countries the use of some of the options is enforced by law, hence no further flexibility is provided;

ii) In areas with serious air pollution problems the use of control technologies may be the only way to site an MEF;

iii) Some control methods reduce one type of emission while increasing another, hence the flexibility provided is a matter of balancing advantages and disadvantages of an option with respect to the potential of the site to absorb different kinds of environmental impact;

iv) Air pollution control may allow for the replacement of old plants in or near urban areas with improvement in ambient concentrations of pollutants;

v) The best option is the use of "clean fuel" (oil desulphurised in the refinery, coal next to the mine) which reduces the emissions without requiring more land or other installations for a power station;

vi) Cooling towers and cooling ponds provide a lot of flexibility in siting large power plants inland.

Siting of Nuclear Power Stations

With respect to the siting of nuclear power stations the Report describes the development of siting policy towards the concentration of generating capacity in fewer sites. There is a danger that such a trend may lead to excessive concentration which may have adverse environmental effects exceeding the economic and safety advantages it may provide and the options offered by underground and offshore siting.

"Cogeneration" of Heat and Electricity

The integrated production of electricity and heat "cogeneration" for both district heating systems and process heat for industry raises the overall efficiency of a power station from about 35-40% to about 85%. In this way less pollutants are emitted for the same amount of useful energy produced and thermal pollution of rivers or sea is largely eliminated. This technology has been used so far only by few OECD countries and to a limited extent. This is due to economic and institutional barriers.

The economic viability of combined heat and electricity production depends on many site specific factors such as size and variation of heat demand, fuel costs, capital investment for district heating networks, interest rates, etc.

The institutional barriers to the adoption of this method are to be found in the co-operation and co-ordination of the plant owner - who may or may not be the electricity utility - and the users of heat and/or electricity. Existing institutions and legislation do not allow in many instances an equal distribution of costs and benefits from such schemes among groups of interest involved. Another difficulty, in case of process heat is for industry to accept dependence for its process steam on an external source.

Although both economic and institutional barriers are not easy to remove, they are not unsurmountable if the political will exists to use this very promising method.

Cogeneration plants facilitate their siting, particularly if they are small (200 MWe and less). They can be sited within urban areas or within industrial areas. They do not need large cooling water resources and produce much less emissions for the same amount of energy. If they burn clean fuel this increases their flexibility in siting much more.

Part II. Socio-economic Impacts

The socio-economic impacts of MEFs on their surroundings are :

i) Increase in population by the influx of a large construction force;

ii) Change in the composition of the local community:
iii) Change in the local labour market;
iv) Pressure on - or inadequacy of - housing, schools, transport systems, public transport, hospitals, social services, recreational facilities, and entertainment;
v) Increase in local business activity;
vi) Increase in income from local taxation;
vii) Further industrialisation.

The similarity of these impacts, independent of the type of MEF, is due to the fact that all these facilities require a long construction period, a big construction site, and a large construction force. Then, once construction is over, their operation requires a much smaller but more specialised work force. Few among other capital intensive industrial facilities can be compared with MEFs in terms of size, investments and construction work force.

The Report examines in some detail the socio-economic effects associated with the siting of nuclear power stations. (Employment, investment, and salaries paid by some MEFs in a number of OECD countries, taxes paid to local authorities by the developers, etc.)

The Report concludes that the siting of an MEF enhances, in general, the economic development of a community. Even when there are strong environmental and economic (local-traditional industry) interests which oppose the siting, and even when there is the possibility that the balance and influence of local political groups may be affected, there are grounds for bargain and compromise among developer, local authority and local public. What is however indispensable is that the local authority participates in the planning and is given the financial means in advance to provide the necessary infrastructure for coping with the increased population.

SITING POLICIES AND PROCEDURES

The participant groups in the siting process are identified (developers, local authorities, environment protection groups, national and regional government) and their roles described as they were traditionally cast. Developers and local authorities were by far the most important.

In the last 10-15 years many important changes took place. On the technical side, many more MEFs were needed, their individual size grew enormously (1) and new technologies appeared. On the socio-political side, Member countries desired to achieve a greater degree of security in energy supplies, environmental and consumer

1) e.g. power plants from 100-250 MWe to 500-1,000 MWe refineries from 10-50 thousand barrels throughput to 100-300 thousand.

movements appeared, and an emphasis on improved quality of life became apparent.

Under the pressure of these changes OECD governments realised that some new policies and regulatory procedures ought to be adopted to accommodate the large number of interested groups and to evaluate the impacts of MEFs on a wider basis. The Report describes a number of these developments both at the national (federal) and the regional level of government.

In the light of these developments and of the analysis of the problems as presented in Chapters I and II, four basic principles for regulatory procedures dealing with the siting of MEFs are proposed:

i) The planning for MEFs should be based on long term policies integrated to national and international energy policies.

ii) The siting of MEFs and the assessment of their environmental impacts should be carried out within long term land use plans.

iii) Regional authorities should indertake the major responsibility in the siting procedures.

iv) Public participation should be encouraged and incorporated in all stages of the siting process.

The application of these principles may be ensured by the adoption of a framework for a procedure for siting MEFs. Such a framework should ensure that policy decisions taken at the national level are pursued in a compatible way by regional and local authorities as increasingly site specific decisions are being reached. Thus the siting of MEFs may be considered as a three stage process where :

a) The first stage (responsibility being mainly of the national or federal government) seeks consensus agreements over long term energy policies so that they are not opposed by interested parties or the public, and may ensure that the siting of MEFs will not act as a constraint to those energy policies.

b) The second stage (responsibility being mainly with regional authorities) seeks to identify and secure, within long term land use plans, possible sites for those types of MEFs which will be necessary for the implementation of long term energy policies, to evaluate the potential of the physical and social environment to absorb the expected impacts and to seek the acceptance by the public of such MEFs.

c) The third stage (responsibility shared between regional, local authorities and the developer) seeks to ensure that once a particular MEF is needed the site which will be selected is in accordance with long term land use plans, that the social, economic and environmental impacts of the technology used at that particular moment in time are evaluated, that the developer and the local authority provide the necessary infrastructure to reduce adverse social impacts, and that the developer is assisted with the licensing procedures so that the project is not delayed.

Guidelines For Siting MEFs

Before proposing guidelines for the siting of MEFs we present the recommendations of the OECD Council which were forwarded to Member governments [C(76)162, Final] as a result of the work of the Task Force on Energy and the Environment and which are relevant to the siting of MEFs, of the work on transfrontier pollution [C(74)224], and of the work on coastal management [C(76)161]. Also are relevant recommendations among those proposed [C(77)109, Final] for the reduction of environmental impacts from energy use in the household and commercial sectors.

a) General Recommendations

i) Environment policies and energy policies are integrated, both at the formulation stage and the implementation stage;

ii) The public is objectively informed and its views are sought;

iii) Land use planning is employed which takes into account environmental protection goals;

iv) Consideration is given as to differences in costs and benefits to those directly affected and to the nation as a whole when examining energy production and use;

v) Energy conservation measures which have positive environmental effects should be promoted.

b) Recommendations on the Siting of Power Plants

i) Solutions acceptable to the interested parties are actively sought within the siting decision process;

ii) Legislative or administrative means be found to encourage the development of siting policies at national level as part of energy development and environmental policies;

iii) Electricity and heat utilities be encouraged, whenever appropriate, to become combined electricity and heat producers, subject to all relevant environmental protection regulations;

iv) Industrial users be encouraged wherever appropriate and

where this will lead to improved economic use of resources, to: (a) increase the "inhouse" generated proportion of their total energy requirements and (b) to market surplus energy, and that these measures comply with environmental controls and standards and be coordinated with electricity and heat utilities;

v) Within each country there be a system to assess the environmental impacts of energy facilities (including comparison with those of other industrial developments) either by preparing EISs or by other comprehensive assessment methods.

c) <u>From "household and commercial sectors"</u>

i) The energy distribution system and the utilisation of clean fuels in high density urban population areas should be progressively improved to meet environmental requirements.

ii) Land use planning for urban areas should formally incorporate an evaluation of environmentally desirable energy systems such as district heating, and of urban designs which would lead to the reduction of the environmental impact of energy use.

d) <u>From Transfrontier Pollution</u>

Countries should define a concerted long-term policy for the protection and improvement of the environment in zones liable to be affected by transfrontier pollution.

Without prejudice to their rights and obligations under international law and in accordance with their responsibility under Principle 21 of the Stockholm Declaration, countries should seek, as far as possible, an equitable balance of their rights and obligations as regards the zones concerned by transfrontier pollution.

e) <u>From Coastal Zone Management</u>

i) Defensive planning, consisting of restrictions, should be complemented by positive planning indicating where activities may be located provided that due consideration is given to environmental protection;

ii) The potential impact on the coastline of significant public and private projects should be assessed prior to their development;

iii) The public should be informed of facts and plans relating to coastal development and involved in the planning process at the earliest possible stage;

iv) The protection of the most aesthetic, culturally and/or environmentally vulnerable areas should be given special care and kept for those activities which, by their kind and scale, are compatible with the preservation of the characteristics of these areas. In addition, areas representative of particular natural systems should be preserved for future study and to serve as regenerative centres;

v) The siting of industrial activities which have to be located in coastal areas should be such as to guarantee a maximum of environmental protection.

Recognising the need for improving the quality of the environment, and for securing sites for MEFs on a long term basis, and in the light of the above Council Recommendations, of the information and analysis presented in the Report and, of the constitutional, administrative and economic constraints for each particular Member country, the following guidelines for the siting of MEFs have been adopted.

Member governments should:

i) Establish long term land use planning systems on either national and/or regional level and ensure that the siting of MEFs takes place within the framework of such plans according to a set of selection criteria with respect to environmental constraints, as well as economic, agricultural, recreational and sociological aspects;

ii) Ensure that regional land use plans are developed on the basis of, among other factors, environmental assessments of the capacity of the region, or of areas within the region, to absorb the expected impacts from industrial and other development without excessive or irreversible damage to the environment;

iii) Ensure that probable sites for MEFs are identified in land use plans by cooperation among national, regional, local authorities and developers in accordance with long term economic and energy growth projections. The possible environmental impacts from such developments should be identified on a national and regional basis;

iv) Develop, given the constitutional and administrative constraints, the ways and means by which sites for MEFs may be secured for long periods of time without undue loss to property owners, developers or the national economy;

v) Allocate to regional authorities increased responsibilities for the siting of MEFs, and provide them with - or allow them to acquire - the necessary legal, administrative and financial means which will enable them to carry out their responsibilities;

vi) Establish a mechanism acceptable to the parties involved which will allow for the advance payment to local authorities of a part of the expected tax income from the operation of the MEF, to provide the necessary infrastructure in time for the accommodation and provision of communal services;

vii) Ensure that whenever local authorities have the constitutional right to decide on the siting of MEFs they do so by:

   a) accepting among proposed sites only those that are in accordance with the specifications of the long term land use plans;

   b) assessing those environmental and social impacts which concern the local community and which have not been assessed during the preparation of the land use plans;

   c) acting as a coordinator in obtaining for the developer the multitude of permits needed for building and operating the MEF to simplify and accelerate the regulatory process.

viii) Promote contacts between administrative bodies responsible for national energy policy and federations of industry, labour unions, environmental groups and other interested groups;

ix) Develop a system of advance notification, information and consultation with bordering countries for such cases of siting MEFs which are expected to have transfrontier effects, with a view to reaching mutually satisfactory solutions;

x) Evaluate the advantages and drawbacks offered by small energy facilities in environmental, economic and efficiency terms and, if conditions are identified where the advantages are substantial, encourage the development of technologies that can best be utilised on a small scale and the design and operation of smaller units for current technologies.

xi) Evaluate the extent to which the overall benefits for the national economy resulting from the adoption of combined electricity/heat generation systems allow for the provision of economic incentives for a more equitable distribution of benefits and disbenefits among groups of interests involved and for the removal of institutional and administrative barriers to its application.

Chapter I

SITING OF MAJOR ENERGY FACILITIES - STATEMENT OF THE PROBLEM

A. INTRODUCTION - STRUCTURE OF THE REPORT

The availability of energy in all its forms at a reasonable price is an essential and much treasured service of both society and industry. During the last decade, however, the siting of Major Energy Facilities (MEFs), which were indispensable for energy production and transformation, became increasingly difficult in OECD countries.

Obstacles to the siting of nuclear power stations have been met in France, the Federal Republic of Germany, Sweden, Switzerland, the United States and other OECD countries. An oil refinery at Nigg Point in Scotland has been delayed considerably due to environmental and land acquisition difficulties. Planning permission was subsequently granted and a master plan for the refinery agreed. A 700 MWe bituminous coal fired power station in Duisburg (FRG)(1), has been objected to on environment grounds and work is suspended pending further appeals in the Federal Administrative Court. A large coal fired power station (about 4,000 MWe capacity) in the state of Utah in the United States has been cancelled, at least partially, due to objections on environmental grounds. These are a few examples of a phenomenon which is to be found in most OECD countries with respect to siting of MEFs and other large, polluting, industrial installations (cement factories, steel mills, etc.).

These unusual features did not appear overnight, but were the result of a sequence of events developing slowly in the 1950s and at an accelerated rate in the 1960s. Unfortunately it did not obtain due consideration on the part of either governments or industry at that time.

The siting of MEFs, as indeed of any other major industrial installation, has been guided for a very long time largely by technical and economic criteria. Industry developed traditionally next to the sources of its raw materials or to urban and communication centres - on the sea coast or along major waterways. Unsettled national economic situations - poverty, unemployment and the regular

1) One of a planned series aiming at a total of 6,000 MWe by 1980.

advent of wars, with the destruction that followed - ensured that industrial installations with their large tax base and employment opportunities were welcome by local and regional communities.

In the post Second World War period OECD countries' industrial growth reached record levels. The construction of industrial plants which took place in the 1950s was followed by an enormous expansion of industrial production which was fuelled by the increase of world trade and the availability of cheap petroleum. Based on economies of scale and conditions of full employment the main trends for industrial plants in the 1960's were automation and gigantism (refineries, chemical plants, steel mills, power stations, etc.).

The result of this expansion was a large increase in heavily industrialised, densely populated zones along the sea coast or large waterways. As per capita incomes increased, families began to own one or often two motor cars and many of them could even afford a second "weekend" house, usually chosen in areas of high amenity value. Furthermore, with their increased mobility, people had the freedom to select where to settle, and consequently tended to object to changes of the environment they had chosen. Correspondingly, as the basic needs of a large part of OECD citizens were taken care of (i.e. per capita income reached a generally satisfactory level), the desirability of economic growth and industrialisation per se began to be questioned. People became concerned with "job satisfaction", "quality of life" and "protection of the environment".

These new concerns found their expression in the establishment of environmental interest groups (national, regional, local and ad hoc) which, by acting as lobbyists or pressure groups, helped to politicise the issue of environmental degradation.

Governments reacted to this movement by passing legislation regulating air and water emissions and by establishing ministries for the environment. Environmental protection thus became part of the thinking of administrations, regional and local governments, the public at large and industry. Industry found that it not only had to abide by existing legislation, but that environmental abuse could produce serious public relations problems. Thus, in the 1970's the situation is such that while the siting of an MEF may be accepted by a part of the population, primarily due to their favourable economic impacts for a given area, another part of the population may be strenuously opposed, believing that the adverse environmental and social implications nullify the advantages.

Unfortunately, the equilibrium between considerations of economic development on the one hand and "quality of life" considerations on the other (including environmental protection) is a tenuous one. The difficulties with the siting of MEFs derive from the fact that it is a process which lies at the interface of these two concerns.

In summary, the opposition to the siting of MEFs in OECD countries may be attributed to the following:

a) The high rates of industrialisation and the high density of population (national and regional);
b) The necessity for most MEFs to be sited on coastal or riverine zones which, in many instances, are heavily industrialised and populated;
c) The very large size of individual MEFs;
d) The effects of their impacts (social, economic, health, environmental) being felt not only by the local community, but by many surrounding communities as well. This includes concern about the consequences of accidental releases of radioactivity from nuclear installations;
e) The objection of a community to dramatic changes in the character of their social and cultural environment;
f) The concern of citizens, and government, with the protection of the environment and the improvement in the quality of life; and,
g) The demands of citizens to participate in decisions which are likely to affect their lives.

Objective of the Report

The objective of this Report is to examine the problems posed by the siting of MEFs and to provide information to Member governments on siting procedures and policies so that the adverse environmental and social impacts of MEFs are reduced while the implementation of national and international energy policies is not jeopardised.

Structure of the Report

The remainder of Chapter I describes (i) the physical and socio-economic impacts that are produced by different types of MEFs, (ii) the extent of the siting problems that OECD countries will be facing in the next twenty years in terms of type and numbers of MEFs and number of sites; and (iii) the way by which these problems developed during the past decade to reach the level of complexity they have attained today.

Chapter II examines some practices which help to reduce the adverse impacts deriving from the siting of MEFs, i.e. to reduce the effects of change to the natural and social environment brought about by an MEF.

The first part deals with physical impacts and in particular with: (i) the flexibility in siting that a developer may have by choosing different types of emission control technologies for fossil fuel power stations, (ii) the alternatives for siting nuclear power plants in terms of remote siting, clustered sites, nuclear power parks underground and offshore siting; and (iii) the advantages of combined heat and electricity generating plants with respect to reduced thermal pollution, energy conservation, and flexibility in siting.

The second part deals with social and economic impacts. These impacts are associated with a large construction site and, later on, with the operation of a capital intensive industry. The overall impacts are largely beneficial and, if advantage is taken of them, they may help to balance the inconvenience of the physical impacts, and enhance the acceptance of an MEF by the local community. Lack of planning, on the other hand, may produce adverse social and economic impacts. Some suggestions are made on how this may be avoided.

Chapter III examines the decision making process for the choice of the sites and in particular the distribution of responsibilities among national, regional and local governments and the regulatory procedures in force in Member countries.

It makes the point that procedures are not just administrative tools, but that they can lead to fewer adverse consequences in siting an MEF. To that end it argues that:

a) siting of MEFs should be an integral part of national long term energy policies;
b) regional government has a more important role to play than in the past;
c) better sites can be chosen within the framework of a long term regional land use plan;
d) the assessment of environmental impacts of individual MEFs should be an integral part of the regulatory procedure.

B. CHARACTERISTICS OF MEFs

An MEF can enhance the economic growth of a region by increasing local tax revenues and providing employment, but it also has the potential for introducing a number of adverse effects including:

Air and water pollution
Noise
Disruption of the ecosystem
Risk of major accidents including radiation effects at nuclear power plants in the event of an accident
Adverse health effects
Adverse aesthetic/amenity effects
Adverse social/economic effects
Conflicts with existing industry.

There is a need for special procedures and security forces to prevent diversion of nuclear materials for non peaceful uses.

The magnitude of these problems for a local community has been exacerbated over the last two decades by the greatly increased scale of MEFs and other industrial facilities. For example, power plants grew from typical sizes of 100-250 MWe to 500-1,000 MWe and refineries increased from 10-50 thousand barrels per day to new plants which may process 100-300 thousand barrels per day.

To better understand the nature of the MEF siting problem it is important first of all to provide brief descriptions of such facilities. Table I.1 summarises, in both quantitative and qualitative terms, some of the important characteristics for a set of "typical" MEFs (1), which will be discussed first for physical impacts and then in terms of their social/economic implications.

a) Physical Impacts

i) Fossil Fuel Power Plants

The first four rows of Table I.1 give information for 1,000 MWe fossil fuel steam power plants, utilising different fuels and different types of environmental controls. Air emission and water use are the most important environmental concerns associated with fossil fuel fired electric power plants. The extent of the problem can be illustrated by the fact that in the United States in 1970, the following percentages of total effluent from combustion processes were due to electric power generation: 50 per cent of SO_x, 25 per cent of NO_x and 25 per cent of particulates. Coal-fired power plants have the potential for the greatest air emissions, but comparison of rows 1 and 2 indicates that particulate removal and SO_x scrubbing can reduce significantly the problem (2). However, it should also be noted that the reduced air pollution in row 2 causes a substantial increase in land requirement, primarily due to the area needed for sludge disposal. Natural gas fired power plants (row 4) are clearly the most attractive plants in environmental terms, but this option is not really available for most OECD countries due to the limited resource base and the preferential use of gas in space heating and

---

1) The quantitative estimates are given to represent "average" or "typical" cases, and these vary from plant to plant depending on design, specific fuel characteristics, level of environmental control, etc.

2) There are, however, some other emission data which indicate that a significant difference of NO_x emissions exists between coal fired and oil fired power plants, and that one should examine carefully the probability of increasing NO_x emissions from coal fired as compared to oil fired power plants.

Table I-1

ENVIRONMENT AND SAFETY CHARACTERISTICS OF MEFs

(Quantities given refer to annual amounts for each typically-sized MEF)

Type of MEF	Air emissions					Water emissions				Land		Safety (b)
	SO_x 10^3 (tons)	NO_x 10^3 (tons)	Part. 10^3 (tons)	Rad. (curies)	Other	Organic 10^2 (tons)	Thermal (10^{12} Kcal)	Rad. (curies)	Water Consumption (10^3 m^3)	Direct use (hectares)	Exclusion Area (hectares)	
A. Electric Power Plants (a)												
1. Coal-fired: No Controls	120	22	48.5	3.10^{-2}(k)	Heavy Metals	Neg.	7.8	0	0	205	203	0
2. Coal-fired: with SO_x scrubbing and wet cooling towers and particulate retention	12	22	2.6	(u)	Heavy Metals	Neg.	0(d)	0	14,000	342	342	0
3. Oil-fired: No Controls	47	21	1.6	5.10^{-4}(k)	0	Neg.	7.8	0	0	59	59	0
4. Gas-fired: No Controls	.02	11	.43	0	0	Neg.	7.8	0	0	33	33	0
5. Nuclear: LWR	0	0	0	5.10^3-10^4(i)	0	0	0(d)	0.1-0.5(l)	24,000	20	314(e)	**
B. Refineries (c)												
1. Uncontrolled	15	12	1.7	0	Smell (mercaptans)	2		0	135,000	521	521	*
2. Controlled	8.7	12	1.7	0	Smell (mercaptans)	.4		0	13,400	521	521	*
C. LNG Import Facilities (f)	.002	.37	.07	0	CO aldehydes	0	0	0	0	410	(u)	**
D. Oil Terminals	(u)	0	(u)	0	0	(u)	0	0	(u)	50	65	*
E. Coal Synthetic Fuel Plant Low BTU gas (g)	3	10	1.4	(u)	Heavy Metals	.01	0(d)	(u)	10,400	164	164	*
F. Liquefaction (h)	1.8	11.8	.5	(u)	Heavy Metals	.004	0(d)	(u)	(u)	240	240	*

a) Calculated from References (1) and (3). Based on 1,000 MWe plant with 75 per cent load factor.

b) ** = very major issue; * = important concern; 0 = not important.

c) From (1) and (2). Based on 300,000 bbl/day (15.5 million tons/year) capacity.

d) Wet cooling tower.

e) Based on Germany's Class I (Typical average value 2 km radius).

f) Based on 250 million cubic feet per day, 7,075 million m^3/day.

g) 1,000 million cubic feet per day.

h) Based on 65,000 bbl/day, 3,35 million tons/year.

i) Mostly noble gases (typical values).

j) Areas indicated are from USA literature. Other OECD countries frequently use one fifth to one tenth of the areas indicated.

k) ^{226}Ra and ^{228}Ra in particulate form (typical average values).

l) Tritium (10-250 curies/y) not included.

cooking. Although not shown in Table I.1, tall stacks can be used to disperse pollutants over a wider area and thereby reduce local air pollution impacts except when inversion occurs. However, there is considerable debate over their use because of the ultimate effects caused by long range transport of pollutants.

The water impacts from electric power plants are caused by their large cooling requirements. Due to the thermodynamics of the process and with current technology approximately 60 per cent of the energy in the fuel is ejected into the environment. The water impacts can take one of two forms depending on the type of cooling system, either: (a) thermal pollution of water with once-through cooling (e.g. rows 1, 3 and 4); or (b) water consumption with cooling ponds or wet cooling towers (e.g. row 2). These water impacts can be eliminated through the use of dry cooling towers, but this option is considerably more expensive, reduces plant energy efficiency by several per cent over wet cooling towers, and creates greater visual appearance problems.

ii) Nuclear Power Plants

Under normal operating conditions, nuclear power plants have less environmental impact on the immediate environment than do fossil fuel plants, the only major disadvantage being the larger waste heat emissions due to the lower efficiency of light water reactors (LWRs (1)) (row 5 of Table I.1). The routine radioactive emissions are considered to be inconsequential at least for the immediate future. However, this does not mean that nuclear power is without environmental problems - there are very serious concerns with (i) the possibility of a major incident (either accidental or from sabotage) that could release large amounts of radioactivity, (ii) radioactive waste disposal. As an illustration of the safety concerns note that direct land use of nuclear power plants is very small compared to fossil fuel plants, but the "exclusion area" (2) is larger than the land needed for these plants.

iii) Refineries

Rows B1 and 2 give physical characteristics for 300,000 bbl/day (15.5 million tons per year) refineries without and with environmental

---

1) The type of nuclear power reactor currently in use by most OECD countries.

2) Exclusion area: area surrounding the plant site where no resident population is allowed for safety-related reasons. In several cases it coincides with the plant fenced area.
Low population zone: area surrounding the exclusion area where the population growth and land use are kept under statutory control for safety-related reasons.

controls, representing typical old and new refineries respectively. Refineries are very large installations (covering 520 hectares) with air and water emissions, large water requirements for processing and cooling - unless air cooling is used extensively - and safety problems due to the risk of explosions and fires. For these typically sized plants, the air emissions are generally smaller than with fossil fuel power plants. However, refineries can emit certain odoriferous gases which, although small in quantity and with no known health impacts in such dilute amounts, can be quite a nuisance to the surroundings.

iv) Liquified Natural Gas (LNG) Import Facilities

Under normal operating conditions, a LNG import facility (consisting of an unloading terminal, storage tanks and a regasification plant) would emit very few pollutants. However, its major disadvantage is the potential for a major accident - an explosion and fire and possibly the release of large amounts of liquified natural gas. There is considerable scientific debate over the consequences of a major LNG spill. Some argue that it would boil off harmlessly into the atmosphere, while others claim it could result in a major explosion or fire. Because of this uncertainty, many argue that LNG facilities should be placed away from major population centres.

v) Oil Terminals

Oil terminals are much simpler installations when compared to the previous and the following. Oil arrives either by a subsea pipeline (offshore oil fields) or by tanker. It is stored in large tanks where it is allowed to settle and separate from gas and brine (or ballast water). Gas treatment plants may be added in the case of offshore oil. Oil is then pumped towards a refinery (through underground pipelines) or to tankers or barges. The landward requirements of the terminal are satisfied relatively easily so that marine access for large tankers (deep sea port) and/or proximity to the resource in case of offshore oil is the main location factor. The most conspicuous feature is the oil tank farm. The area required for storage is, on present experience, .75 hectares per 20,000 tons excluding land used for processing, ballast water storage and treatment and landscaped areas. Taking account of all requirements, a site of about 60-65 hectares is needed to accommodate a terminal with a throughput of about 20 million tons a year. This problem has been met with underground cavern storage (Sweden) or by careful landscaping (BP, Dalmeny, near Edinburgh). The other source of pollution is the return of ballast waters (after settling) in the sea which contain some hydrocarbons (content is regulated by the authorities) as well as the oil which finds its way into the sea as the result of loading and unloading operations.

vi) Coal Synthetic Fuel Plants

Coal synthetic fuel plants (either the production of high-Btu gas or synthetic crude oil) with planned environmental controls would have physical characteristics not unlike coal fired power plants or refineries. However, comparing rows E and F with rows A2 and B2, indicates that the level of air and water emissions and the water and land requirements should all be less for coal synthetic fuel plants. Of course, it should be noted that these figures are based on "typically sized plants" which, for the refinery, is a capacity of 300,000 bbl/day (15.5 million tons per year) but for the coal liquification plant is only 65,000 bbl/day (3.3 million tons per year). Thus, on a per unit output basis, the coal synthetic plant would have greater physical impacts than a refinery.

vii) Hydroelectric Power Generation

The share of electricity generated at hydroelectric power plants has been declining in most OECD countries. The use however of hydroelectricity from "pumped storage", an indirect way of storing electricity has been increasing, and this trend is expected to continue until other low cost methods for storing electricity are developed. At present, "pumped storage" is the only commercial method for storing electricity.

Although in industrialised countries most favourable sites for hydroelectric plants have already been utilised, some still remain to be developed. These are generally, on streams or rivers with moderate flows of water or where the terrain is such that the dams which will be needed would be relatively small. Many of these dams serve also for other purposes, such as flood control and recreation. Unlike other MEFs, hydroelectric plant siting is restricted by the rate of flow of the water along the river and by soil and geologic conditions which allow or not the building of a dam, so that flexibility in the choice of the site is very much limited.

Political, social and environmental factors have been also considered when siting hydroelectric dams. The political and social concerns arise from the objections of the local population to the inundation of towns and villages and the flooding of agricultural, recreational forest or game producing areas. There is also concern that the dams may break either because of construction flows or as a result of earthquakes leading to extensive loss of life, flooding and property damage.

Construction of a dam involves a number of potential adverse environmental effects that are somewhat different in nature from those of other MEFs. They include adverse impacts on migrant and resident species; increases in water temperature due to increased water surface area; degradation of water quality through oxygen depletion; increased production of algae and increase of certain organic and metabolic products. For dams at which there are wide variations in intermittent stream flows, benthic organisms and fish spawning beds can be destroyed, growth of valuable plants may be reduced, and the dilution capacity of the stream for organic wastes can be severely decreased.

viii) Transmission Lines

The trend towards larger and fewer electricity generating plants has an obvious benefit to the environment. However, since these plants can seldom be sited close to load centres because of their potential adverse environmental effects, a network of high voltage transmission lines has to be erected. This network, once in place, acts as a constraint to the choice of sites for power stations which have to be built later on. The reason is both economic and environmental.

Above ground transmission lines are objected to on aesthetic grounds because of their size, large number and the long uninterrupted corridors that are required for their construction. They interfere with radio and television reception. Switch gears and transformers generate noise which is a nuisance particularly at night. Possible adverse health effects may result from the increasing voltage of transmission lines on the public exposed to it.

By careful design of transformers and the use of acoustical enclosures noise can be reduced. Switchgear noise reduction can be accomplished by the use of silencers. The aesthetic effects are eliminated, if transmission lines are placed underground. This method has a much higher cost (10, 15 times that of overhead lines) and unless new technologies are developed to help reduce that cost, underground transmission lines can be used only selectively, when they cross areas of high amenity value.

b) Social and Economic Impacts of MEFs

Although the physical impacts of MEFs vary according to the type of the facility, the social and economic impacts, with the exception of health effects and the special problems associated with public concern over nuclear power are more homogeneous because they derive from large construction projects and from the operation of large, capital intensive industrial installations. If the development has been correctly planned on a long term basis and the financial means have been provided in time to local authorities to meet the anticipated change, adverse impacts can be reduced and the community would enjoy the substantial economic advantages that an MEF brings about.

Two phases can be distinguished: the construction phase and the operation phase. The most severe adverse impacts may be expected during construction.

Construction Phase

The period of construction may last from three to ten years according to the type of the facility. For a 100,000 barrels per day refinery employment requirements are approximately 1,000 man years spread over five years. For a 1,000 MWe nuclear power station employment varies according to country and technology from 3,000 man years to 10,000 man years spread between seven and nine years, i.e. an average of 400 to 1,400 workers per year, the peak being 2,400 workers (see Table IIB1). Based on experience in France about half of this force brings along their families, and thus, assuming an average 2.5 persons per family, a total of approximately 1,100-2,300 new inhabitants have to be accommodated for seven to nine years within commuting distances to the site.

i) Housing

Unless the site is next to an industrial or urban area with some unemployment so that workers already living nearby can commute to the construction site, most of the work force will be new and will have to be accommodated in new dwellings or in caravans/trailers. Experience with nuclear plants in France showed that for a two unit 2,000 MWe project, the need for family dwellings varies between 102 and 272, and for single persons between 124 and 332. About 10-15 per cent of the workers live in provisional accommodation.

ii) Schooling

Again from experience in France, about 200-500 children, half of whom are of school age, move into the area and have to be accommodated in local schools and nurseries. During the seven to nine years of construction the extra classes needed will shift upwards and when the construction force leaves, the community may find itself with excess school capacity.

iii) Infrastructure, social services, recreation

Roads and public transport are two areas that may be found under the greatest pressure from the influx of new workers, but social (e.g. hospitals) and recreation are also important. Again, if the site is close to urban centres, the problems are not as difficult. But in remote areas (e.g. Scotland) some services such as doctors or police may be seriously lacking. Moreover the immigration of a new labour force into a remote and sparsely populated area may alter significantly the composition of local society and produce upheavals in the local way of life.

Operation Phase

MEFs are capital intensive industrial installations and although they need a substantial labour force for operation, this is much smaller than that needed for construction, hence the social requirements are not as large as for the construction phase. For the operation of two 1,000 MWe nuclear power plants, about 200 persons are needed, while for four units, 350 persons are sufficient. An area which has supported the construction activities can accommodate the operation personnel with little difficulty.

The possible adverse social and economic impacts which may result from the sudden drop in population can be reduced provided that local authorities have anticipated them and have been provided with the necessary financial means by the developer or the government to create new jobs for the local labour force which found employment in the construction, or to disperse this force to neighbouring areas.

Economic benefits

MEFs, as indeed any other major industrial installation, brings substantial economic advantages to the locality.

i) Local tax income

Taxes paid by industry to local authorities are calculated on a different basis among OECD countries, e.g. on capital investment, on salaries paid annually, etc. In any case the local tax income from MEFs is very substantial, particularly when these are sited among small communities. For example, in France, an upper limit of 5,000 Francs per year per person was set, the rest being distributed among neighbouring communities and the region as a whole, when it was found that small communities were unable to utilise all the tax income from a new nuclear power station. In the United States, where local tax rates are decided by local authorities, the income can amount to as much as from $6-$15 million annually for a site of 4,400 MWe.

ii) Employment

From the employment figures given previously it is evident - particularly in times of unemployment - that the economic advantages of the construction of an MEF for a community are substantial, even when it provides half of the labour force needed. Although it is true that local industries may suffer from the competition of a better paying employer - in particular for skilled labour - the net result is undoubtedly positive. This is also the case for the operational personnel. When the MEF goes into operation the number of employees falls considerably and varies among different types of installations. A 200,000 barrel per day refinery may employ up to 450 while a 1,000 MWe power station may employ up to 500 employees.

iii) Local Business

An influx of approximately 1,500 to 2,500 persons during construction spending nearly 75 per cent of their income in the community is again a substantial boost to local business. About 50 per cent of that is spent in rent and food.

C. EXTENT OF THE PROBLEM FOR OECD COUNTRIES

To evaluate the MEF siting problems that OECD countries will be facing in the next five years use is made of the projections for energy production and consumption described in the OECD publication "World Energy Outlook" published in 1977 (Ref. 5). Time horizons chosen for this publication are the years 1980 and 1985, and two scenarios are proposed: (i) A Reference Scenario which assumes a 3.6% annual energy growth rate for OECD countries and (ii) an "Accelerated Scenario" which assumes that accelerated development of indigenous energy resources and energy conservation measures will lead, by 1985, to an overall saving in imports of 324 Mtoe compared to the Reference Scenario.

The possible MEF siting problems are examined here in terms of each of the two scenarios. For the Reference scenario it is assumed that, given the lead times needed for building and putting on stream such facilities, sites for the MEFs needed to meet the projections for 1980 have been received while this has not been done for the sites needed for 1985. Following the pattern of the "World Energy Outlook", data based on the Reference Scenario will be presented first and will be followed by those based on the Accelerated Scenario.

a) MEFs for Oil Production and Transport

Table I-2 presents figures for oil production, imports and total consumption of oil in million tons for 1974 (a), 1980 (b) and 1985 (c) for the six OECD "regions" (Canada, U.S.A., OECD Europe, Japan, Australia-New Zealand) according to the Reference Scenario.

The two basic energy policies proposed for OECD countries, namely to reduce oil imports and to develop their indigenous energy

Table I-2

OIL

(Reference Scenario - in million tons)
Reference /5/

	Area		Production	Imports	Total	Increase	New (1) Refineries	New Oil (2) Terminals
1	OECD (Total)	(a)	634	1,266	1,900			
		(b)	810	1,497	2,307	407	4	
		(c)	887	1,750	2,637	330	23	9 - 16
2	Canada	(a)	94	-9	85	-	-	
		(b)	76	38	114	29	1	
		(c)	69	54	123	9	1	0
3	U.S.A.	(a)	498	290	788	-	-	
		(b)	540	458	998	210	1	
		(c)	582	477	1,059	61	6	3 - 6
4	OECD Europe	(a)	22	708	730	-	-	
		(b)	177	624	801	71	0	
		(c)	222	738	960	159	5	3 - 6
5	Japan	(a)	1	263	264	-	-	
		(b)	1.5	348	349.5	85.5	2	
		(c)	3.5	441	444.5	95	10	3 - 4
6	Australia-New Zealand	(a)	20	14	34	-	-	
		(b)	15	30	45	11	0	
		(c)	11	40	51	6	1	0

1) 200,000 barrels per day or 10.3 million tons of crude per year
2) 25.5 million tons oil per year

(a) = 1974, (b) = 1980, (c) = 1985

resources, affect the siting of oil related energy facilities in two ways.

The number of new MEFs is reduced (see, for example, the figures for oil fired power stations in Table I-4); and some new MEFs will be needed to serve the two new OECD oil producing areas, Alaska and the North Sea.

(i) Refineries

Table I-2 presents figures for new refineries on the basis of the projections for oil consumption. Following the 1973 oil price crisis and the slowdown in industrial activity 20-30% of the oil refining capacity in most OECD areas remained idle, particularly during 1974-1975. The figures for new refineries presented in Table I-2 have been arrived at by assuming that:

- an increase of up to 25% of the 1974 oil consumption will be taken care of by existing refining capacity unless there are special identifiable reasons (e.g. North-Sea Oil);
- new refineries will have an average capacity of about 10 million tons per year (200,000 barrels per day).

The figures of Table I-2 indicate that a substantial part of OECD Europe's oil needs will be met by indigenous production (North Sea). It may be expected that two and possibly three new refineries will be built in the United Kingdom and Norway. The oil imports of non EEC European countries will be mostly among Mediterranean countries, and given their relatively small size, a larger number of smaller refineries will be needed than that shown in the Table. In the U.S.A., if a substantial part of Alaskan oil is directed towards the west coast, the refineries indicated in the Table will be shared with the east coast where the bulk of Arabian Gulf and North African imports arrive. In Japan, an increase in refining capacity seems to be necessary to meet the projected increased imports of oil.

(ii) Pipeline terminals and deep sea ports

The North Sea coast is the main OECD area where siting pipeline terminals and deep sea ports will be necessary. It may be possible that the expansion of existing terminals - Tyneside, Dalmeny/Hound (21 million tons per year), Sullom Voe (expected to reach 100 million tons per year) and Flotta (13 million tons per year) will be sufficient, but it seems very probable that deep sea ports and terminals will be developed in other North Sea countries as well.

There are three other areas for increased oil imports: the east coast of the U.S.A. where two deep water ports each of a capacity of 100,000 tons per day seem to be needed, Japan, where the increased imports call for at least three such ports, and the Mediterranean coast where, unless existing facilities are substantially enlarged,

Table I-2A

OIL

(Accelerated Policy Scenario - In million tons)
Reference /5/

			Production	Imports	Total	Increase [1]	Refineries [1]	Oil Terminals [1]
1	OECD (Total)	(a)	634	1,266	1,900	-	-	-
		(b)	1,007	1,217	2,224	324 (737)	5 (23)	4 (11)
2	Canada	(a)	94	-9	85	-	-	-
		(b)	70	35	105	20 (38)	0 (1)	0 (0)
3	U.S.A.	(a)	498	290	788	-	-	-
		(b)	703	211	914	126 (271)	0 (6)	0 (3)
4	OECD Europe	(a)	22	708	730	-	-	-
		(b)	222	554	776	46 (230)	0 (5)	2 (5)
5	Japan	(a)	1	263	264	-	- -	- -
		(b)	3	382	385	121 (180.5)	5 (10)	2 (3)
6	Australia New Zealand	(a)	20	14	34	-	-	-
		(b)	11	35	46	12 (17)	0 (1)	

(a) = 1974, (b) = 1985

1) ( ) Corresponding figures of the Reference Scenario, table I-2.

one or two new terminals will be needed to cope with the oil imported through the enlarged Suez canal and from North Africa.

iii) Accelerated Scenario

Table I-2A presents figures for oil production, imports and total consumption for 1974(a) and for 1985(b) according to the accelerated scenario.

With respect to new refineries there is an important difference between the two scenarios. Only Japan, among OECD regions, will need new refineries. The U.S.A. and OECD Europe will need even less refining capacity than they used in 1974.

With respect to oil terminals fewer units will be needed (4) than in the Reference Scenario (11). This is due to the reduction in imports, but the projected increase in indigenous production will call for new terminals to serve the North Sea and Alaska.

b) MEFs for Natural Gas Production and Transport

Table I-3 presents figures for production, imports and total consumption of natural gas in billion cubic metres for 1974(a), 1980(b) and 1985(c) and for each of the OECD regions according to the Reference Scenario. Changes in the trade of natural gas are expected to lead to a net increase in imports. These changes are due to:

(i) the need to use more of the OPEC or the North Sea gas which was flared in the past;

(ii) the exhaustion of U.S.A. known reserves and the delay in their replacement by Alaskan gas;

(iii) the proximity of the North African gas fields to the Mediterranean coast of OECD countries;

(iv) the Japanese policy of increased import of LNG from the Arabian Gulf.

These activities will lead to the need for a substantial number of LNG terminals, storage tank capacity and pipeline capacity.

(i) LNG Terminals

The figures in Table I-3 show the probable need for new LNG terminals. These figures were based on the following assumptions:

An "average" modern LNG tanker has a capacity of 100,000 tons/dwt and carries about 46 million cubic metres of natural gas. An "average" LNG terminal gasifies and transports about 2.5 billion cubic metres of gas per year, i.e. one tanker per week. Larger than "average" tankers have been built (125,000 tons dwt) and even larger are planned (160,000 tons dwt). Also much larger LNG terminals are planned as, for example, the second FOS terminal in France which is

Table I-3

NATURAL GAS

(Reference Scenario - In billion cubic metres)
Reference /5/

	Area		Production	Imports	Total	Increase	New LNG (1) Terminals
1	OECD (Total)	(a)	818	15	833	-	-
		(b)	816	66	882	49	-
		(c)	909	132	1041	159	15-20
2	Canada	(a)	68	-26	42	-	-
		(b)	80	-25	55	7	0
		(c)	98	-35	63	8	0
3	U.S.A.	(a)	587	24	611		
		(b)	492	37	529	-8	
		(c)	539	46	585	56	3-4
4	OECD Europe	(a)	155	11	166	-	-
		(b)	227	29	256	90	
		(c)	230	82	312	56	3-5
5	Japan	(a)	3	6	9	-	
		(b)	5	27	32	23	
		(c)	10	47	57	25	8-10
6	Australia New Zealand	(a)	5	-	5	-	
		(b)	13	-	13	8	
		(c)	32	8	40	27	1

(a) = 1974, (b) = 1980, (c) = 1985

1 Mtoe = .85 billion m3 gas

(1) 5 billion m^3 gas per year

expected to treat 35 billion cubic metres of gas per year, i.e. one 100,000 LNG tanker per day. The figures in Table I-3 assume that new LNG terminals will have an average capacity of 5 million cubic metres per year.

It can be deduced from the figures in Table I-3 that Canadian imports will not be sufficient to cover the U.S.A. needs for 1980 and 1985. The extra 8 billion cubic metres which will probably be imported from the Arabian Gulf and North Africa call for 3-4 "average" LNG terminals which will be sited on the east coast and the Gulf of Mexico, unless Indonesian gas imports call for one of these to be sited on the west coast. There is still an unknown but important factor namely the time of completion and operation of the Mexico-U.S.A. gas pipeline.

For OECD Europe besides the advent of North Sea gas, imports of North African gas by Mediterranean countries will be substantial. Assuming that the 8 billion cubic metres from Eastern Europe (Ref. 4, Annex A, Table A25, p. 101) find their way to Western Europe by pipeline, an extra 10 billion cubic metres should be imported by 1980 and another 40 billion by 1985, probably from North Africa. If the FOS project comes on stream in time, only another three to five "average" LNG terminals will be needed and will probably be sited along the Mediterranean coast.

Lastly, increased imports of 21 billion m^3 for 1980 and of another 25 billion m^3 for 1985 are projected for Japan. The gas will probably come from the Middle East, Indonesia and, by 1980, from Australia and New Zealand. These volumes call for about 10 LNG terminals (5 billion m^3 per year each) or of fewer terminals of higher capacity.

(ii) The Accelerated Scenario

Table I-3A presents figures for natural gas production, imports and total consumption for 1974(a) and for 1985(b) according to the accelerated scenario.

There is an important difference between the two scenarios with respect to the indigenous gas production. The U.S.A. is expected to produce 85 billion m^3 more than in the case of the Reference Scenario, while OECD Europe, an extra 31 billion m^3. On the other hand imports are projected to increase only by 12 billion m^3. As most of this indigenous gas will be transported by pipeline, only western Europe among OECD regions will need new LNG terminals to deal with the increased trade in natural gas.

c) MEFs for Electricity Generation

Table I-4 presents the electricity generating capacity in OECD "regions" for 1974(a), 1980(b), 1985(c) according to the fuel used

Table I-3A

NATURAL GAS

(Accelerated Scenario - In billion cubic metres)
Reference /5/

	Area		Production	Imports	Total	Increase	New LNG Terminals
1	OECD (Total)	(a)	818	15	833		
		(b)	1,024	144	1,168	335 (208)	14 - 19
2	Canada	(a)	68	-26	42	-	
		(b)	98	-35	63	21 (15)	0
3	U.S.A.	(a)	587	24	611	-	
		(b)	624	47	671	60 (-26)	3 - 4
4	OECD Europe	(a)	155	11	166	-	
		(b)	261	82	343	177 (146)	4 - 6
5	Japan	(a)	3	6	9		
		(b)	10	59	69	60 (48)	7 - 9
6	Australia New Zealand	(a)	5	-	5		
		(b)	32	-9	23	18 (35)	0

Table I-4

ELECTRICITY GENERATING STATIONS

(in GWe, 1,000 MWe nuclear plants, 600 MWe fuel plants)

	Area		COAL			OIL			GAS			NUCLEAR		
			GWe	NEW GWe	NEW PLANT	GWe	NEW GWe	NEW PLANT	GWe	NEW GWe	NEW PLANT	GWe	NEW GWe	NEW PLANT
1	Total OECD	(a)	303			177			105			38		
		(b)	375	54	90	204	27	45	120	15	25	143	105	105
		(c)	452	95	158	244	40	67	117	-3	-5	306	163	163
2	Canada	(a)	7			2			3			2.5		
		(b)	11	4	6	4	2	3	2	-1	-1	5.5	3	3
		(c)	13	2	3	4	0	0	2	0	0	9.8	4.4	5
3	U.S.A.	(a)	186			66			73			19.4		
		(b)	220	34	57	68	2	3	67	-6	-10	71.0	51.6	52
		(c)	294	74	123	67	-1	-2	56	-11	-20	161.0	90	90
4	EEC	(a)	84			56			25			10.0		
		(b)	87	3	5	63	7	12	28	3	5	37.3	26.5	26
		(c)	90	3	5	97	34	57	28	0	-	77.3	40	40
5	Non EEC OECD Europe	(a)	8			11			1			3		
		(b)	15	7	12	19	8	13	5	4	7	15	12	12
		(c)	17	2	3	31	12	20	1	-4	-7	27	12	12
6	Japan	(a)	7			42			3			3.2		
		(b)	6	-1	-2	50	8	13	19	16	27	14	10.8	11
		(c)	12	6	10	45	-5	-9	25	6	10	32.3	18.3	18
7	Australia New Zealand	(a)	12			1			1					
		(b)												
		(c)												

(coal-oil-gas-nuclear) as projected in Ref. 5 for the Reference Scenario. The figures have been calculated from tables A1-A9, pp. 90-94 of Annex A to Ref. 5. Furthermore, it was assumed that new fossil fuel power plants will have an average capacity of 600 MWe and that two such units will be sited together, while nuclear power plants will have an average capacity of 1,000 MWe and will be sited in groups of three units. Hydroelectric plants have not been included because projections in terms of capacities and sites are difficult to be made. Given the lead times needed for building and putting on stream power stations it is understood that the sites for the new plants projected for 1980 have been found and cleared by the regulatory agencies. But for most of the plants projected for 1985 the sites, even when identified, have not been cleared yet by the authorities.

From the data of Table I-4 the following trends are manifest:

i) Coal Fired Power Stations

All new non nuclear power plants in the U.S.A. will be coal fired. New coal fired electricity generating capacity will be also added in Europe (mainly the United Kingdom and Germany) and Japan (Canadian, U.S.A., and Australian coal).

ii) Oil Fired Power Stations

There is a clear trend towards reduction of oil fired electricity generating capacity in all OECD regions particularly in the U.S.A. Only OECD Europe plans considerable new oil fired capacity.

iii) Natural Gas Fired Power Stations

No new gas fired electricity generating capacity is planned for any OECD region with the exception of Japan. On the contrary, there is a trend towards decommissioning gas fired power plants particularly in the U.S.A. The grouping of OECD countries in regions throws however some shade on national policies. Thus, Italy plans an expansion of its gas fired capacity in the 1980s but this does not show in the Table because of the opposite policies of all other European countries.

iv) Nuclear Power Stations and the Role of Nuclear Power in OECD Countries

Nuclear power will take up a much higher proportion of electricity generating capacity in all OECD regions except for Australia and New Zealand. It is a fact that for many OECD countries over the next 20-30 years nuclear power is their only real alternative to imported oil. This dependence may diminish later on if, in the meantime, OECD governments commit substantial resources on new promising forms of energy, and if energy conservation programmes are pursued vigorously and meet with public acceptance.

In spite of this need to rely on nuclear power the programmes of some countries, particularly the European OECD countries are now substantially smaller than they were in earlier forecasts. For example the EEC had set in 1974 a target of 200 GWe capacity for 1985 while the EEC countries themselves accepted a maximum of 160 GWe. The estimate given in the present (1976) projections for the EEC is only 91.4 GWe.

The reasons for these downward revisions are (Ref. 4) the following:

i) The increase in capital costs for nuclear power stations due to increases for construction materials, to high interest rates, to increases in lead times, and to additional expenses to meet safety requirements. The price of uranium has also increased from $7 per pound to $16 per pound. The need to exploit poorer grades of uranium minerals will further increase the price of nuclear fuel and the cost of nuclear power;

ii) The reduction in demand for electricity which followed the 1975 economic recession;

iii) The public anxiety over safety and environmental preoccupations which caused the lengthening of planning and consent procedures and by weakening the commitment of governments and of electricity utilities to nuclear power.

Tables I-5 and I-6 present some more data from different sources to throw some light as to the trends and developments in the number and size of nuclear power stations, the number of sites which will be needed to house the projected capacities and the concentration of generating capacity on each site.

Table I-5 presents some detailed data on the number and size of nuclear power stations and on the number of sites in most OECD countries with important nuclear energy programmes. Figures in columns I, II and III have been obtained from Nuclear Engineering International (NEI) 1977. Figures in columns IV and V were derived by simple calculations from the previous three columns. Figures in columns VIa and VIb were obtained from the OECD "World Energy Outlook" (p. 51 Ref. 4) and those in VIc and VId calculated by subtracting figures of column IVc from those of VIa and VIb respectively. Columns marked A refer to nuclear power stations operating in 1977. These include a large number of experimental stations hence their relatively large number. Those marked B refer to nuclear power stations in construction in 1977. And those under C are nuclear power stations which are being planned in 1977. For the interpretation of the

Table I-5

THE SITING OF NUCLEAR POWER STATIONS IN OECD MEMBER COUNTRIES (1960 - 1985)

Country	I Nuclear Power Stations			II Sites			III New Sites			IV Total Capacity (GWe)			V Average Capacity per site (MWe)			VI OECD Forecast (GWe)		VI Discrepancy OECD - NEI		VII New Sites Needed to meet 1985 Forecasts		
	A	B	C	A	B	C	A	B	C	A	B	C	A	B	C	Ref.[a] Case	Acc.[b] Case	Ref. Case	Acc. Case	NEI	OECD Ref.	OECD Ref.
1 Canada	10	11	4	6	6	7	6	0	1	4.9	12.5	15.6	490	2100	2230	12.8	12.8	-2.8	-2.8	1	0	0
2 U.S.A.	55	116	209	41	69	109	41.5	28	40	41.5	65.9	213.5	1012	1558	1959	52	152	-61.5	-61.5	40	10	10
3 United Kingdom	32	38	42	14	16	17	14	2	1	5.4	9.4	13.4	617	800	896	11.0	13.0	-4.2	-2.2	1	0	0
4 France	7	22	32	5	8	11	5	3	3	3.6	17.8	28.9	724	2234	2627	31.0	35.0	+2.1	+6.1	3	4	6
5 Germany	7	18	29	7	15	20	7	8	5	3.3	15.1	28.0	476	1010	1401	31.0	38.0	+3.0	+10.0	5	7	11
6 Japan	11	27	33	7	14	16	7	6	3	7.1	19.0	24.1	1023	1362	1510	35.0	49.0	+11	+25	3	10	17
7 Belgium	2	3	5	2	2	2	2	0	0	1.3	1.7	3.6	665	870	1795	4.9	5.6	+1.3	+2	0	1	1
8 Sweden	5	10	12	3	4	5	3	1	0	3.3	7.7	9.9	1100	1925	1980	7.8	9.0	-2.1	-0.9	0	0	0
9 Switzerland	3	6	7	2	6	7	2	4	1	1.0	3.7	4.9	527	626	699	3.3	3.8	-1.6	1.1	1	0	0
10 Italy	4	4	8	4	4	6	4	0	2	1.4	1.4	3.8	349	349	633	6.4	7.4	+2.6	+3.6	2	5	6
11 Spain	3	15	18	3	10	11	3	7	1	1.1	8.8	11.9	379	889	1080	14.0	20.2	+2.1	+8.3	1	3	7
12 Other	4	8	9	4	6	7	4	2	1	0.9	3.2	4.5	230	530	640					1		
Total	143	278	408	98	160	218	98	61	58	87.9	169.4	363.9	633	1200	1450	309.0	345.8			58	40	56

A = In operation (1977); B = In construction (1977); C = Planned (1977).

a = Reference scenario

b = Accelerated scenario

figures in the Table two assumptions were made :

i) that sites under C although identified (NEI gives the name of each site) have not yet been cleared by the regulatory authorities so they are considered new sites;

ii) that construction of a nuclear power plant takes 7-8 years to accomplish, hence plants under C will be operating in 1985 and, therefore, figures under C and nuclear power forecasts of OECD for 1985 can be compared.

Table I-6

NUCLEAR POWER STATIONS AND SITES IN OECD COUNTRIES

OECD Country	GWe end 1977 *	Sites 1977**	GWe end 1985*	Sites 1985**
Canada	3.3	3	10	4
U.S.A.	48	48	115	97
United Kingdom	5.4	15	9.4	18
France	6	7	34	15
F.R. Germany	10	11	25	21
Japan	10	9	27	20
Belgium	1.7	2	3.5	3
Sweden	3.2	3	7.4	4
Switzerland	1	2	2.8	4
Italy	0.6	4	5.4	9
Spain	1.1	4	15	12
Others +	0.9	3	3.4	5
Total	91	111	258	212

\+ Netherlands, Austria and Finland.

* Estimates from Forthcoming NEA/IAEA Publication, "Uranium Resources, Production and Demand".

** Taken from IAEA publication, "Power Reactors in Member States, 1977 edition".

Under these assumptions there is a rather good agreement between OECD and NEI figures. The discrepancies in the figures for the U.S.A. and Japan are explained to considerable extent in "World Energy Outlook" (p. 44 for the U.S.A., and p. 57 for Japan).

Column VII gives approximate figures for the number of new sites which will be needed for meeting the capacity projected for 1985. They have been obtained by assuming that each new site will house the average capacity for each country as shown in column IVc. On the basis of this assumption - which, if not completely realistic, is supported by the trend towards housing more capacity per site [compare figures under IV(A), IV(B), IV(C)] - a total of 56 new sites will be needed for 1985 according to NEI, 40 new sites according to the OECD Reference Scenario and 58 new sites according to the OECD accelerated scenario for 1985.

Table I-6 gives some estimates which are based on two other sources: the forthcoming NEA/IAEA Publication "Uranium resources, Production and Demand" and the IAEA Publication "Power Reactors in Member States, 1977 edition". It may be observed that the current III nuclear sites (98 in Table I-5) have an average capacity of .8 GWe (.633 GWe in Table I-5). There is work in progress in 101 sites which will bring the total capacity by 1985 to at least 258 GWe (this figure falls halfway between the totals under columns IVB and IVC in Table I-5 due probably to the difference in lead times of the ongoing construction projects). For this increment the average installed capacity will increase to 1.7 GWe per site (1.45 GWe in Table I-5). It is reasonable to expect as much as 4 GWe per new site between 1985-90. Thus for an additional 198 GWe to come on stream during that period about 50 additional sites will be needed in the whole OECD area (58, 40 or 56 according to Table I-5).

From the above discussion two points may be retained:

i) there is no shortage of projections as to the extent of the OECD commitment to nuclear power in the next 10-15 years;

ii) despite the divergences among these projections, the number of new sites for nuclear power plants which must be identified and cleared in the next five years within OECD countries is between 50 and 60.

v) <u>The Accelerated Scenario</u>

Table I-7 presents the corresponding figures of the Reference and Accelerated Scenarios for 1985. With respect to coal fired power plants there is no significant difference between the two options. This indicates that the development of coal resources in OECD countries is considered an essential policy and the limits set in the Reference Scenario cannot be exceeded.

The big difference is, as can be expected, in oil fired power stations. Among OECD regions only OECD Europe is expected to add some further capacity, while in all others not only new capacity will not be built, but also many old plants will be decommissioned, thus presenting an overall reduction in the number of sites.

With respect to gas fired power stations most areas do not show any substantial difference apart from the U.S.A. where the assumed deregulation of natural gas prices is expected to lead to an increase in output. Thus the trend towards decommissioning of plants indicated in the Reference Scenario changes to a trend towards adding some new capacity.

Table I-7

COMPARISON BETWEEN REFERENCE AND ACCELERATED SCENARIOS FOR ELECTRICITY GENERATING STATIONS (1985)

	Area		Coal				Oil				Gas				Nuclear			
			GWe	New GWe	New Plant	New Sites	GWe	New GWe	New Plant	New Sites	GWe	New GWe	New Plant	New Sites	GWe	New GWe	New Plant	New Sites
1	OECD Total	R	452	149	248	124	244	67	112	56	117	12	20	10	329.6	288.6	289	97
		A	435	188	222	111	165	25	-5	-21	134	29.1	49	24	370.8	330.4	331	111
2	Canada	R	13	6	10	5	4	2	4	2	2	-1	-2	-2	8.7	6.1	6	2
		A	13.1	6.1	10	5	.4	-1.6	-3	-3	1.6	-1.4	-2	-2	8.7	6.1	6	2
3	U.S.A.	R	294	108	180	90	67	1	2	1	56	-17	-28	-28	173.6	152.8	153	51
		A	284.5	98.5	164	82	44.9	-11.1	-19	-19	77.7	4.7	8	4	173.6	152.8	153	51
4	OECD Europe	R	107	15	25	13	128	61	102	51	29	3	5	3	112.6	98.4	99	33
		A	115.81	23.8	40	20	86.5	19.5	33	17	29.5	3.5	6	3	139.8	126.0	126	42
5	Japan	R	12	5	9	5	45		5	3	25	22	37	19	34.8	31.4	32	11
		A	11.5	4.5	8	4	32.7		-16	-16	25.3	22.3	37	19	48.7	45.5	46	16
6	Australia/New Zealand	R																

Table I-8

POPULATION DENSITY, ENERGY CONSUMPTION DENSITY OF OECD COUNTRIES, URBAN AREAS AND UNUSABLE LAND

	Country	Population Density (persons/ 1,000 km^2)	Energy * Consumption Density (Toe/ 1,000 km^2)	Urban ** areas as % of total	Unusable ** land as % of total
1	Canada	2	17	-	-
2	U.S.A.	23	186	4.2	16.0
3	Japan	295	900	4.8	-
4	Australia	2	8	-	-
5	New Zealand	11	36	0.9	23.1
6	Austria	90	275	1.1	13.9
7	Belgium	320	1,488	26.7	1.3
8	Denmark	117	413	10.8	9.2
9	Finland	14	66	3.1	32.9
10	France	96	325	4.3	4.7
11	Germany	250	1,063	9.3	6.6
12	Greece	68	97	-	-
13	Iceland	2	11	-	-
14	Ireland	44	106	-	-
15	Italy	184	454	-	-
16	Luxembourg	137	1,907	-	-
17	Netherlands	367	1,670	9.2	21.8
18	Norway	12	59	2.2	69.9
19	Portugal	96	86	-	-
20	Spain	69	119	-	-
21	Sweden	18	98	0.9	38.2
22	Switzerland	156	535	4.3	21.5
23	Turkey	50	23	-	-
24	United Kingdom	230	897	12.0	4.0

* Figures are for 1975.

** Figures obtained from OECD document AGR/WP1(75)4 Rev.

With respect to nuclear power notable changes are expected in OECD Europe and Japan.

d) Other Factors which Influence the Siting of MEFs

Table I-8 shows some of the other factors that appear to be important in the future availability of sites. These are population density (persons per 1,000 km^2), energy density (tons oil equivalent, toe, per 1,000 km^2), urban area as percentage of the total land area, and unusable land again as percentage of the total land area.

Although the rate of population growth in OECD countries is low when compared to that of many other geographical regions, some OECD countries are so densely populated that relatively large numbers of people are exposed to the adverse effects of MEFs. Related to this is the energy density (tons oil equivalent per 1,000 km^2), which gives a rough idea of the concentration of the adverse impacts associated with energy production and consumption. As population and per capita energy consumption increase more and more people will necessarily suffer the adverse impacts associated with MEFs (unless the adverse impacts are limited to some extent with pollution control technologies and proper planning), and thus the siting problem is likely to become more difficult.

Of course, these "country average" figures on population and energy density do not give the whole picture for three reasons. First, for many OECD countries they vary significantly from one region to another. Secondly, in all OECD countries the majority of the population is concentrated in urban areas that represent a relatively small proportion of the total land area. Because of this concentration of population and the fact that for economic and energy efficiency reasons MEFs are usually located near the end users, MEFs tend to affect larger segments of the population than would be indicated by simply examining national average figures. And finally, even when MEFs can be sited in sparsely populated areas, experience in a number of countries (e.g. Scottish experience with North Sea oil and gas development, and the United States experience with coal and oil shale development in western United States) indicates that siting is still likely to be a very difficult problem. As will be discussed later, public opposition in such instances can be explained by large relative changes that are introduced in the local area.

Water is also an important physical factor affecting the availability of sites for MEFs. Many MEFs, for technical, economic or energy efficiency reasons must be located along the sea coast or on large lakes or rivers. Countries with no, or only a short, sea coast would appear to face more difficulties than others. However, siting along the sea coast can create resistance because large percentages of the population live along the coast and because of potential conflicts with other valuable uses such as recreation and fishing. In terms of fresh water, total requirements for MEFs represents a constraint for few OECD countries on a national basis, but there are numerous examples of localised instances where water use is a major concern (e.g. in the western United States and along the Rhine River).

D. CAUSES FOR THE OPPOSITION AGAINST THE SITING OF MEFs IN OECD COUNTRIES AND THEIR HISTORICAL DEVELOPMENT

a) Environmental Movement and Public Participation

After World War II a number of social protest movements emerged, particularly in the U.S.A., e.g. the "Ban the Bomb", antiwar, civil rights and black liberation movements which had far reaching implications for the development of the environmental movement. The particular issues of the early protesters such as social and political inequality, poverty, and the "credibility gap" broadened in scope and began to question the foundations of social thought and of institutionalized arrangements.

In the meantime, large scale technological development swept the western world. Motorways, river dams, synthetic fertilizers and pesticides, petrochemicals, synthetic fibres and plastics, etc., epitomized a technology based on replaceable "throwaway" items. But the innovations - high nitrogen fertilizers, long lived pesticides, throwaway containers, etc. - had harmful unintended effects. Abuse of the environment and the diminishing stock of resources could no longer be ignored. The environmental movement gained momentum in the late 1960s through the cumulative effect of these issues. It gained exposure through the mass media and literature which drew public attention to the salient aspects of environmental degradation.

The mounting tension due to alienation from governmental decision making led the public to demand a more direct role in the many phases and levels of the government decision-making process. The concerns of the public were converted into effective political and social responses primarily through public interest organisations. These organisations encourage a broad spectrum of organised groupings that, at times, emphasize environmental concerns. It is difficult to distinguish the role of environmentalists in the public decision making process from that of other political activities.

The specific goals and motivations of public participation cannot be easily characterized. The "participatory process" reflects those concerns of the public that are not effectively represented in the political forum. The public strives to establish effective outlets for information and for the expression of its views. The disillusionment and disappointment of the public on issues of environmental policy in the late 1960s led to a distrust of official decision-making and opened the regulatory agencies and the governments' policy making process to criticism and scrutiny by the public.

The planning and construction of energy facilities employ resources in many instances which are in direct competition with recreational, aesthetic or other uses and values. The desire to preserve such environmental amenities has led the public to become actively involved. Often, concerns other than potential environmental degradation have led the public to participate in revisions on the siting of MEFs. This is the case of the nuclear power issues.

b) Concerns Over Nuclear Power

Broader concerns which are associated with a large expansion of nuclear power electricity supply can be divided into three major issues: (i) radioactive waste disposal; (ii) fuel reprocessing; (iii) safeguards and nuclear proliferation.

i) Radioactive Waste Disposal *

The operation of nuclear reactors and other nuclear facilities produces radioactive by-products of many kinds. Some wastes contain small amounts of short-lived radionuclides and can be disposed of immediately in the environment. Some other wastes which contain longer-lived radionuclides in greater quantities and which are produced in gaseous, liquid and solid form must be treated and conditioned before storage and/or disposal. These wastes, which are usually referred to as low-and-medium-level wastes can be disposed of by shallow land burial, by emplacement in deep excavations within suitable geological formations and by sea dumping.

At present, the management of radioactive wastes takes place under the supervision of national regulatory bodies; however, concern is occasionally expressed about the increase in waste production expected to result from the expansion of nuclear programmes. It is conceivable that in the future the current procedures for the management of low-and-medium-level waste may require modification.

Of greater radiological significance than the above are the high-level wastes which are produced during the reprocessing of spent fuel. With the present controversy over fuel reprocessing particularly in the U.S., it has been suggested that the spent fuel itself

* The NEA Committee on Radiation Protection and Public Health charged a group of experts with the task of preparing a report on radioactive waste management. The report, entitled "Objectives, Concepts and Strategies for the Management of Radioactive Waste Arising from Nuclear Power Programmes" was published in September 1977. This report examines in depth waste management issues. A short description of the problem is presented here to put into perspective the concern of part of the public which sometimes leads to opposition to nuclear power in general or to specific nuclear plants.

might be considered as waste and disposed of in appropriate geological formations after a suitable storage period. Regardless of the option eventually adopted and of the nature of the final products requiring disposal, it is certain that the back end of the nuclear fuel cycle will produce extremely hazardous materials that need to be kept isolated from the biosphere during many thousands of years.

There are a variety of management options being considered for high-level waste and/or spent fuel. As far as high-level waste is concerned, the most promising strategy is based on the following steps:

- storage as liquid;
- vitrification;
- storage as a solid;
- disposal into deep geological formations either on land or under the ocean floor.

The volume of high-level waste resulting from vitrification of the liquids produced by reprocessing the fuel irradiated in a 1000 MWe LWR in a year is about 3 m^3. Many processes have been proposed for the solidification of high-level waste. However, most of them have not yet been demonstrated on a fully operational scale. On the other hand, significant experience has been gained with engineering-scale pilot plants, particularly in France and the U.S. A full-scale vitrification plant called AVM (Atelier de Vitrification de Marcoule) is planned to start operation in France in the spring of 1978.

With respect to the disposal of solidified high-level waste, several Member countries have launched geologic disposal programmes. By far the greatest effort is spent on geological formations on land, but five Member countries have decided to co-ordinate their activities and to explore jointly the option of waste emplacement into the argillaceous sediments underlying the deep ocean floor. A good deal more is known about disposal in salt formations than about the other rock types. This is due to the programmes in the U.S. and the Federal Republic of Germany where salt has been under consideration for about twenty years.

In the Federal Republic of Germany, the Asse salt mine has been used for several years as a disposal facility for low-and-medium-level waste and for experiments related to the emplacement of high-level waste in salt. The safety analysis work performed so far indicates that there are many geologic environments capable of containing long-lived waste with the greatest reliability, which may alleviate the anxiety of part of the public with respect to nuclear energy.

A common question related to waste disposal is the probable cost of the operation. The report mentioned above includes an annex on financing of waste management which reviews current cost estimates for the various waste management operations. The cost of waste disposal is estimated at about 1% of the total cost of nuclear electricity generation. However, other recent estimates are somewhat higher and without doubt there is a wide range of uncertainty; nevertheless, there is agreement among specialists that the cost of waste disposal cannot increase to the point where the overall economics of nuclear energy would be significantly affected.

In spite of the fact that geologic disposal appears to be an adequate solution for long-lived radioactive waste, there remains opposition to the creation of long-lived radioactive materials which will retain their toxicity for many thousands of years. An additional argument is that the lack of operational disposal facilities in deep geological formations makes the concept unproven. However, since the operation of geologic disposal facilities for several years or even a few decades would not prove the successful containment of the waste over many thousands of years, the only possible demonstration of the adequacy of geologic disposal schemes is conceptual in nature and consists of showing that the risks of all hypothetical containment failure events are acceptably low. This conceptual demonstration is being provided by the safety analyses of geologic disposal systems.

It would be desirable that such demonstration work be undertaken based not only on theoretical studies on the long-term integrity of potential disposal formations, but also on in situ experiments which would provide the necessary data base for realistic safety assessments. Such work would undoubtedly benefit from close international co-operation between all countries active in this field as this would also contribute to better public acceptance*.

ii) Fuel Reprocessing

Fuel reprocessing techniques were established early in the history of nuclear energy, when plutonium had to be extracted from the fuel of reactors for the manufacture of nuclear weapons. At the early stages of the nuclear electricity programmes reprocessing was not considered a priority, but as the number of nuclear power stations increased some OECD countries have built fuel reprocessing facilities (notably United Kingdom, France, U.S.A., Federal Republic of Germany). However, the only commercially operating facility at present is a 800 ton/year plant in France. Such facilities are planned for operation in 1985 in a number of other countries (such as Federal Republic of Germany, United Kingdom, Japan, Belgium). As a result of the recent U.S. policy to prevent nuclear weapons proliferation; there are no plans for reprocessing operations at present in the U.S.A.

The reprocessing of used fuel from LWRs, the type of reactor in prominent use today, has several advantages such as the reduced need

* Such co-operation is already being promoted by the OECD Nuclear Energy Agency through its Radioactive Waste Management Committee.

for uranium (thus reducing the requirements for mining, milling and enrichment), and alleviating to considerable extent the waste disposal problem by recycling the plutonium in reactors. Reference [3] indicates that the "throw away" option results in 100% of the plutonium produced appearing as waste material instead of 1 to 2% which is, at present, the estimated loss during the various steps of an industrialized plutonium fuel cycle. For other advanced reactor types such as the HTGR and the FBR reprocessing of spent fuel is a necessity. Reprocessing generates several waste streams each of which has to be treated in a different way. For some of them disposal facilities on an industrial scale is a necessity. Until these facilities are provided, there will be a need for storage, firstly on the site of production and possibly also in centralized facilities specially designed for this purpose. In addition, storage may also be required in production sites before the waste is conditioned into its final form. Thus the disadvantages of reprocessing compared to the throw away option are: the greater amount of plutonium and other high radioactive wastes handling the increased security risks, the dangers of nuclear proliferation and the safeguards which have to be adopted to minimize such risks.

iii) Safeguards and Nuclear Proliferation

Safeguards refer to the procedures and management arrangements used to control and account for radioactive materials (primarily uranium and plutonium) in the nuclear fuel cycle. The objectives are to maintain accountability and to prevent inadvertant losses or diversion for non-peaceful uses of these radioactive materials. Closely related is the issue of widespread proliferation of nuclear materials and technology that could be used to manufacture atomic bombs. LWR fuel is low enriched uranium which cannot be used for explosive devices. But the prospect of plutonium recycle with LWRs or breeder reactors has focussed attention on the safeguards and nuclear proliferation issues because plutonium can be used to manufacture atomic bombs.

Some have argued that adequate physical protection of nuclear materials may call for a security force so substantial as to threaten personal liberties and that regardless of present or future arrangements, radioactive materials, especially plutonium, may be accidentally lost or diverted for non-peaceful purposes. In response to concerns for nuclear proliferation, a number of countries modified previous attitudes regarding plutonium. For example, in late 1976 the U.S. announced plans for a moratorium on fuel reprocessing and a ban on the sale of nuclear fuel reprocessing equipment to other countries. This implies a delay of at least several years of the decision on plutonium recycle for LWRs and of the development of FBRs. Since this time the subject of fuel reprocessing and plutonium recycle has been the subject of intense international consultations, but no firm agreements have been reached.

It should also be noted that an International Nuclear Fuel Cycle Evaluation Programme (INFCE) has recently been undertaken at the initiative of the United States, with the participation of some 40 countries. In this evaluation, where all issues pertaining to different nuclear fuel cycle alternatives will be reviewed, the need to prevent proliferation of nuclear explosive devices will be the most significant factor.

The technical complexity of the above subjects has prevented a broad public understanding of many of the issues, especially those pertaining to nuclear risks. The debate has sometimes been confused by the intervention of extremist opponents. However, many concerned citizens and some prominent scientists have questioned, in an objective manner, the wisdom of nuclear power development. This has had a very important effect in challenging the technical expertise of responsible administrations and in demonstrating that ultimate decisions should rest on sound technical facts but may be influenced by other political and socio-economic factors. This has enabled both sides to use scientific information selectively and has carried the debate to a higher level of technical complexity, thereby putting some issues even further beyond the comprehension of the layman.

Another interesting aspect of the opposition to nuclear power is that, rather than being a unified movement with common interests, it is comprised of coalitions of widely different interest groups - some are only opposed to nuclear plants at particular sites, others just want improved designs and tighter controls, while others want a complete ban on any type of nuclear power development. The coalition nature of the opposition, combined with the broad and shifting set of concerns mentioned previously, contribute to the lack of clear overall objectives and the elusive character of the opposition to nuclear power.

Of course, it is possible that the opposition will die away as public attention is captured by other concerns, as people become accustomed to living with nuclear risks, and perhaps under the pressure of real or prospective energy shortages. However, the more likely case is that the development of more technically informed opposition groups and the internationalisation of the debate will give continuing expression to the anxieties. Policy makers, therefore, cannot afford to ignore the existence of public opposition, nor should they accept the view which ascribes the opposition only to ill informed extremists.

c) Opposition to the Siting of MEFs

MEFs are large installations which require much space and are among the most polluting industrial plants. As noted previously, most MEFs are to be found in areas which offer themselves to industrial and urban development, i.e. along the coastline, on the banks of major rivers and the shores of lakes. Due to (a) the increase in the size of electric power plants and other energy processing facilities; (b) the increasing demand for fresh water in agriculture, industry, urban and recreational areas; (c) the need for extensive transport systems and other infrastructure; and (d) the fact that MEFs become nuclei for further industrial and urban development, the siting of MEFs has become a problem of the coastal and river zones, par excellence. The use of technical and economic criteria alone for choosing the sites led to the neglect of social values which were not as evident or not as loud, as economic growth, and had to be abused to a considerable extent to provoke a reaction on the part of the public.

Until recently the siting of MEFs, apart from nuclear power stations, was largely decided between the developer and the local authority which had the major responsibility for drawing and implementing land use plans. It was very much an administrative procedure with little, if any, public participation. An MEF was a source of taxation income for the local authority, of jobs for the local labour force, of increased business for the local community, and of possible further industrial development. By the existence of a local electric power plant and in the absence of a strongly connected national grid the community was guaranteed the availability of electricity at a reasonable price.

For as long as "... the prosperity and well being was counted by the number of stacks that could be seen from one's window..." such policies held well. But when excessive concentration of industry in OECD countries led to high per capita incomes and high educational standards, and to considerable deterioration of the environment, awareness of the potential conflicts between industrial development and the quality of life became widespread, and the criteria of well being had to change.

The first indication of this concern on the part of citizens regarding MEFs appeared with the advent of nuclear energy which had the disadvantage of being initiated in the shadow of the development of nuclear weapons and the hydrogen bomb. As early as the late 1950s, in the United States, the public began to oppose the installation of nuclear electricity generating units because of fear of radioactive fallout - a characteristic of nuclear weapon experiments which were carried out at that time - of the novelty of the technology, and of radioactivity emissions both in routine operations and in case of an accident. The fact that the United States Atomic Energy Commission (A.E.C.) was both the developer and the regulatory authority for nuclear development added to the scepticism of the public as to the extent and accuracy of the evaluation of hazards associated with nuclear energy.

An examination of a number of these early cases of protest shows that they followed a very similar pattern:

i) the protest movement was begun by a strongly motivated, usually eloquent individual;

ii) a local group of citizens was created to enlist the support of the community and bring pressure on local politicians and the A.E.C.;

iii) the local group obtained scientific and technical advice;

iv) the local group sued in the Courts;

v) local and national new media gave extensive coverage to the debate;

vi) local politicians and political parties moved in to make political capital out of the issue;

vii) regional and national environmental protection groups (many of them newly established) took up the issue and dealt with it on a national policy level;

viii) national environmental groups began to play the role of the individual activist whenever the siting of a new nuclear power plant was announced.

There were three main developments in the domain of environmental protection from the siting of MEFs during the 1960s and 1970s which follow the pattern described above for the opposition to nuclear power, namely:

a) The institutionalisation of the opposition, from the committed individual to the national environmental group;

b) The involvement of a large number of groups (developers, land owners, communities, local authorities, administration departments, environmental groups, labour unions, etc.);

c) The politicisation of the issues, with the involvement of local and national politicians and of political parties.

The end result of these developments was a much more complex set of procedures (e.g. applications, permits, environmental impact statements, public hearings, etc.) for decisions that were previously almost entirely in the domain of the private sector.

Another important result of these early developments was that both "sides" adopted an attitude of confrontation when dealing with the issues. Thus, a very large percentage of cases were taken to the Courts just because the relevant Administration department refused to prepare an Environmental Impact Statement. The public and environmental groups have, on the other side, been refusing MEF developments for which better sites could not be proposed. This attitude of confrontation, which is very much alive today, has helped to aggravate, or even to create, many of the problems associated with the siting of MEFs to the detriment of a country's interest, both economic and environmental.

In the late 1960s, early 1970s environmental protection legislation dealing with air and water quality was passed in most OECD countries. The establishment of emission, and in some cases, ambient concentration standards affected siting decisions only with regard to sparing highly industrialised areas with ambient concentrations of pollutants close to the maximum acceptable standards, and in some cases requiring the use of pollution control devices, e.g. high stacks, electrostatic precipitators, cooling towers, etc.

But at that time three new developments appeared to complicate the issues involved.

a) The increase in the size of MEFs made their impacts felt farther than their immediate physical and social environment thus giving them regional, even national significance. Benefits and costs became more important, and it was more evident that they were disproportionately distributed among the population. The issue could not be treated by developer and local government alone, as the impacts surpassed the latter's jurisdiction.

b) The public did not object only to emissions exceeding ambient concentration or emission standards but also to change of the environment, be it social, economic or physical.

Thus opposition to siting MEFs in agricultural areas has been expressed by local population (1), where neither amenity value of the land was high nor ambient concentration standards of pollutants were to be exceeded. On the other hand, MEFs were welcomed in heavily industrialised, heavily polluted areas, particularly in times of local unemployment.

c) The 1973 oil price crisis indicated that OECD governments had to examine carefully, and take important decisions, with respect to their energy policies. The need to develop indigenous energy resources to replace imports, the large nuclear electricity programmes of some OECD countries, the slow progress towards adopting energy conservation measures, and the inertia towards shifting considerable energy R & D resources to new, less polluting energy resources, have indicated that the siting of MEFs may become even more complicated, unless improved measures are taken to deal with the problems involved.

It is in the light of these new developments that this Report is written.

1) Examples of this are the coal power plants in North-Western United States, the nuclear power plant at Baud et Saint-Louis near Bordeaux, the deep water port at Sullom Voe in the Shetlands, etc.

Chapter II

MITIGATING THE ADVERSE IMPACTS OF MEFs

Chapter I pointed out that the opposition to the siting of MEFs can generally be explained by the adverse changes that such a facility can bring, whether the changes are of a physical, social or economic nature. If this is the case, then the constraints on siting could be relaxed by any measures which reduce the adverse changes. This Chapter will examine a number of such measures, first of a technical kind and then of a social/economic kind, in order to analyse to what extent they can give increased flexibility in the siting of the MEFs.

Part I

THE TECHNICAL APPROACH: THE CASE OF ELECTRIC POWER PLANTS

As discussed in Chapter I, MEFs can produce a number of adverse physical changes to the environment, including:

- air and water pollution;
- noise;
- disruption of the ecosystem;
- potential for major accidents;
- adverse visual/amenity effects.

For any given MEF there are technical options for reducing such adverse physical effects, but usually at some cost in either economic or efficiency terms, or both. Of course, there is usually no way to eliminate entirely the physical impacts associated with an MEF. The issue is how much economic or energy efficiency should be sacrified to reduce the adverse physical effects.

The purpose of this section is to examine some technological approaches for reducing the physical impacts of MEFs, the economic costs involved and their potential effects on siting. However, since it would not be feasible to cover the subject in depth for all types of MEFs, the topic will be illustrated by considering the case for electric power plants. Four categories of technologies will be considered: (1) air pollution controls on fossil fuel power plants; (2) reducing the adverse effects of siting nuclear power plants; (3) cooling systems for power plants, and (4) combined heat and electricity generation. For each category, first the technologies will be described in terms of their environmental, economic and energy efficiency characteristics and their current stage of development. Following this will be an analysis of the extent to which these technologies can give increased flexibility in siting.

A. AIR POLLUTION FROM FOSSIL FUEL POWER PLANTS

Much attention is being paid, both by electric utilities and regulatory agencies, to the environmental effects of electricity generation. The main problem is air pollution arising from the emissions of dust particles and oxides of sulphur and nitrogen. There is little doubt that air emissions from fossil fuel power plants are one of the major factors underlying public concerns that have led to obstacles in the siting of such plants and to substantial R & D programmes to develop technologies for reducing the air pollutants. Table II-A1 lists some information on emissions, efficiency, and economic costs for 1,000 MWe plants employing different types of fuels and different types of air pollution controls. It should be recognised that these data are given to represent "typical" cases and that actual values can vary substantially among plants depending on many factors such as heat value of the coal, type of cooling system, labour costs, interest costs, etc. This is especially the case for the economic cost estimates, and for this reason economics will be discussed in more detail after the following brief description of the information in Table II-A1.

The first row of Table II-A1 gives figures for a conventional plant with no emission controls, burning an "average" coal - 6,650 Kcals per kilogram, 2.6 per cent sulphur and 12.5 per cent ash. As indicated in the table such a plant emits to the atmosphere 190,000 tonnes annually of SO_x, NO_x and particulates. The solid waste is ash collected from the boiler and due to its large volume it can sometimes represent a serious disposal problem. Where possible the ash is sold for use in the manufacture of light-weight building materials, road building, or land reclamation. Where commercial opportunities do not exist, it is stored in waste piles. The nature of regulations in individual countries for handling these solid wastes determines whether the piles are unsightly and whether they create air and water pollution problems. Option 1 represents a "base case" and would probably not be acceptable in many OECD countries - either because of local public concerns with the air pollutants or because of national pollution standards or both. Option 2 is for a plant employing particulate retention systems and tall stacks -

both of which are in widespread use in many OECD countries. Particulate retention removes 99 per cent of the particulate emissions and although the NO_x and SO_x emission rates are not reduced, the tall stacks reduce ground level concentrations of these pollutants. Whereas the other measures considered relate to reducing emissions, the purpose of tall stacks is to disperse pollutants into the atmosphere as efficiently as possible so as to avoid excessive ground level concentrations downwind of the power station. High stacks may, of course, be utilised with measures to reduce the quantity of emissions: the one is not a substitute for the other. The efficiency of dispersion, or the avoidance of "plume droop" depends not only on the height of the stack but also on the temperature and velocity at which the flue gas is released to the atmosphere. Some additional cost is incurred both for increasing the velocity and for increasing the temperature of the flue gas, but the effects on electricity costs are very small. Thus, for a relatively small incremental cost over Option 1, some local concerns with air pollution may be reduced.

Sulphur occurs in coal in two forms: as organic sulphur chemically bound to the hydrocarbon constituents and as inorganic pyritic and sulphate sulphur in the minerals associated with the coal. Options 3 and 4 illustrate cases for cleaning of coal to remove the sulphur prior to combustion at plants that in most cases would be located near the coal mine. Option 3 represents a typical case using <u>physical</u> cleaning and particulate retention (at the power plant). Physical cleaning can remove up to 80 per cent of the inorganic (pyritic) sulphur and a significant portion of the ash, but it has no effect on the organic sulphur. Thus the level it can reduce SO_x emissions varies greatly depending on the particular coal characteristics. It has been estimated in the United States that 13.5 per cent of the reserves could be physically cleaned to meet SO_x emission standards of 1.2 lb. of SO_2 per 10^6 BTUs. The costs to pay for physical cleaning are approximately a 10 per cent loss in total system efficiency (incurred at the cleaning plant) and about a 7 per cent increase in electricity costs. Physical cleaning technology is commercially available but it is used in some OECD countries to remove sulphur for coal used in power plants. Option 4 represents the case of <u>chemical cleaning</u> to produce a "clean fuel". Chemical cleaning removes both inorganic and organic-bound sulphur, but this technique is currently in its development stages and thus is not in commercial use. Option 4 assumes approximately 90 per cent of the sulphur and 95 per cent of the ash is removed so that SO_x and particulate emissions would be greatly reduced with no controls required at the power plant. It is estimated this option will cause about a 15 per cent loss in system energy efficiency and increase the costs over the base case by about 19 per cent - but such estimates should be

Table II-A1

TYPICAL CHARACTERISTICS OF 1,000 MWe FOSSIL FUEL POWER PLANTS (75 PER CENT LOAD FACTOR)

	Options	System efficiency (per cent) (i)	Annual air emissions (10^3 tonnes)			Annual solid waste (10^3)	Percentage increase electricity costs (j)
			SO_x	NO_x	Particulates		
1	Conventional coal-fired boiler; no controls; average coals (a)	40	120	22	48	300	-
2	Conventional coal-fired boiler; particulate retentions and tall stacks; average coal (b)	40	120	22	3	345	0.5
3	Conventional boiler mechanically cleaned coal; particulate retention (c)	36	60	22	2	345	6
4	Conventional boiler; chemical cleaning; particulate retention (d)	34	12	22	1	50	40
5	Coal-fired boiler; FGD (throwaway) (e)	37	12	22	1	1,000	10
6	Coal-fired boiler; FGD (regenerable) (e)	35	12	22	1	350	13
7	Atmospheric fluidised bed (throwaway) (f), average coal	39	12	10	0.5	1,000	-4
8	Pressurised fluidised bed (regenerable); average coal (f)	38	12	5	0.5	350	13 (k)
9	Oil-fired boiler; average residual oil; no control (g)	40	74	21	2	0	-
10	Oil-fired boiler; desulphurised residual oil; no controls (h)	37	15	21		0	13

(a) Average coal assumes 6,650 Kcals per kilogramme; 2.6 per cent sulphur; 12.5 per cent ash.

(b) Assumes 99 per cent effectiveness in particulate retention.

(c) Assumes 50 per cent of total sulphur and some ash is removed in coal cleaning and 99 per cent effectiveness of particulate retention.

(d) Assumes 90 per cent of sulphur and 95 per cent of ash is removed.

(e) Assumes 90 per cent SO_x removal efficiency and "average" coal.

(f) Assumes 90 per cent SO_x removal and lower NO_x emission. Based on average coal.

(g) Based on 2.5 per cent residual oil.

(h) Sulphur reduced from 2.5 per cent to 0.5 per cent with 8 per cent energy lost and a cost of $23 per tonne.

(i) System efficiency includes the conversion efficiency of the power plant and the efficiency of the fuel cleaning (if any).

(j) Percentage increases for options, 2 through 8, are in relation to option 1; increase for option 10 is in relation to option 9.

(k) See the two footnotes on page 60.

References: [6] and [7].

regarded with caution due to the very early state of development of chemical coal cleaning.

Flue gas desulphurisation (FGD) is a post-combustion technique for removing SO_2 from combustion gases. It works by passing the combustion gases over or through a material that reacts with the SO_2. Options 5 and 6 give typical data for non-regenerable (throwaway) and regenerable FGD systems, respectively. Both types of systems are commercially available. Regenerable systems have only been tested on a few small power plants. Many experts still consider the technologies, especially the regenerable systems, to be unproven and their reliability and cost uncertain. Both would usually be designed to reduce SO_x emissions by about 90 per cent. The throwaway systems would do so at a cost increase of about 10 per cent (including disposal costs), would reduce generation efficiency by about 8 per cent and would create a very large solid waste disposal problem - approximately 1 million tonnes annually of sludge (for the plant size given in Table II-A1), which for the limestone processes, is a mixture of fly ash and water plus calcium sulphite and calcium sulphate. The regenerable flue gas desulphurisation systems have the advantage of eliminating this disposal problem by regenerating and re-using the reactive compound in the sorbent. These processes recover SO_2 from the combustion gases and convert it into marketable by-products such as elemental sulphur, sulphuric acid, or concentrated SO_2 gas. However, they create additional water pollution problems with an additional loss of efficiency and at an additional economic cost (compared to throwaway systems) of nearly 3 per cent (after by-product credits).

Most of the attention on reducing air pollution emissions from stationary sources has concentrated on sulphur oxides and particulates. However, another pollutant, nitrogen oxides are also produced during the combustion of fossil fuels, with approximately 5 per cent to 15 per cent of the total being emitted from electric generating plants.

Control of nitrogen oxides can be accomplished by reducing or preventing the formation of NO_x or by removing the NO_x from the flue gases after it is formed. In addition to the use of lower nitrogen containing fuels, the methods that have been tested to reduce the formation of NO_x involve combustion modifications. These include reducing the rate of heat release during combustion, low excess air operation, flue gas recirculation, the use of steam and water injection to reduce flame temperatures, new types of burners and two stage combustion. Some of these methods are in use today in some OECD countries.

Extraction of NO_x from stack gases has been tested using a variety of scrubbing agents, relative reduction of NO_x by ammonia or other reducing agents and catalytic oxidation using various catalysts. None of these methods are as yet being used commercially.

Like FGD, fluidised bed systems deal with the air pollution problems at the power plant site for both sulphur and nitrogen oxides. Option 7 is for an atmospheric pressure, throwaway fluidised bed boiler, while Option 8 represents a more advanced pressurised, regenerable system (1). Although the atmospheric system is further developed than the pressurised one, neither are commercially available at the current time and most projections estimate they will not be until the early or mid-1980s, for atmospheric systems and the late 1980s for pressurised systems. As indicated by the figures, fluidised bed systems have one inherent advantage over all other options in terms of lower NO_x emissions. Atmospheric, throwaway systems are projected (at least by some experts) to cost less than conventional boilers (because they have high heat transfer rates and thus are smaller), but like throwaway FGD systems they create large amounts of solid waste. However, the solid waste from FGD systems is a wet sludge, while the fluidised bed wastes are a dry solid that should create less of a disposal problem. The pressurised regenerable system creates only ash as a solid waste and yield even lower NO_x emissions, but at some loss in overall efficiency (due to the regenerative process) and at a cost of approximately 13 per cent above the Option 1 base case (2). The cost estimates for both atmospheric and pressurised systems are highly uncertain, however, and it will take several years of development efforts before the economics can be estimated more firmly.

The last two rows of Table II-A1 give data for uncontrolled oil-fired plants burning residual oil. Option 9 is for 2.5 per cent sulphur, while Option 10 is for the case of residual oil desulphurisation from 2.5 per cent to a level of 0.5 per cent. Currently residual oil desulphurisation is being used by some OECD countries and the technology is well known. As indicated in the Table, it is estimated that Option 10 reduces SO_x emissions by a factor of five but with an efficiency penalty of about 8 per cent (incurred at the refinery) and a cost increase for the electricity of 13 per cent. As mentioned previously, comparing the economic costs for various pollution control strategies is risky because the costs can be very sensitive to the specific conditions of the application (e.g. load factor, fuel characteristics, land cost, etc.). This is particularly true for new technologies which are still in the development stage.

---

1) There are regenerable, atmospheric systems and throwaway, pressurised systems possible, but Options 7 and 8 have been chosen just to illustrate the range of characteristics.

2) Without regeneration, the pressurised fluidised bed is expected to have slightly higher energy efficiencies and approximately the same economic costs as the atmospheric systems.

Having reviewed the environmental, efficiency and economic characteristics of several air pollution control alternatives for fossil fuel power plants, the possible effects that these options can have on siting will now be addressed. If physical impacts are a major cause of opposition to MEFs, it follows that reducing those physical impacts would mitigate the opposition. If coal-fired power plants were relatively benign buildings similar in character to a large warehouse, then it is hard to imagine their arousing great concern on the part of the public. It would therefore seem that any of the various options for reducing air pollution would offer a means for alleviating at least some of the constraints attached to the siting of a fossil fuel power plant. Unfortunately, however, the situation is not that simple.

First of all, in many countries there are regulations that limit the emission levels of certain air pollutants; thus, only in the broadest sense do pollution controls give any added flexibility in the selection of sites. Without them no sites are available.

The real concern of the public is with the "impacts", i.e. increased ground level concentrations of pollutants, not just with emission rates. The ambient concentrations will depend not only on the rate of emissions and the stack height but also on the meteorological and topographic conditions of the site and the other sources of pollution in the region. In areas with already serious air pollution problems, the use of some type of air pollution control may allow sites to be developed that would otherwise not be permitted. Experience has shown that there are cases where coal fired plants with no environmental controls other than the use of particulate retention and high stacks, may be considered acceptable by the population in the surrounding area. However, the use of high stacks to reduce ground level concentrations while keeping emission rates unchanged is now becoming clouded by concerns with long-range transport of air pollutants, so that attention is broadened much beyond the immediate vicinity of an MEF.

Another complicating factor in the analysis is due to the fact that some of the air pollution control options discussed earlier can increase other problems that affect siting. One example of this is the flue gas desulphurisation and the fluidised bed system based on "throwaway" processes. While they reduce SO_x emissions by approximately 90 per cent, they create a huge solid waste disposal problem. A throwaway FGD system for a 1000 MWe plant would create enough sludge each year to cover an area of three - six hectares to a depth of ten metres. Thus, even though air pollution is reduced, land requirements for solid waste disposal and its potential for water pollution (both surface waters and groundwaters) could actually limit the number of possible sites rather than helping to alleviate the siting problem.

Finally, it is important to note that in many cases today, power companies are choosing, because of air pollution concerns, sites for power plants in sparsely populated areas away from the major electricity consuming areas. This can cause other environmentally related problems to be exacerbated; for example, the increased transmission distances cause lower system efficiency which means more coal must be burned per unit of final consumption and more land impact and visual amenity difficulties are created due to the high voltage power lines. Therefore, depending on regulations, meteorological conditions and other factors, if air pollution controls would allow plants to be built closer to the major consuming areas, then not only total air emissions but also some other environmental impacts would be reduced. This may be important because certain pollution control techniques might enable electric utilities to redevelop existing urban power station sites by replacing obsolete plants with new, more efficient plants. There is little doubt that the best option from a siting point of view would be the use of a "clean fuel", as it would reduce the emissions without requiring additional land for solid waste disposal (as with "throwaway" systems) or for buildings and wastewater treatment (as with "regenerative" systems). The precombustion cleaning of coal will require siting the cleaning plant, but since this would usually be located at the mine, the incremental problems should not be severe. The desulphurisation of residual oil would only require expanding the capabilities of refineries and thus would not create an additional siting problem. Overall, compared with the alternative of building relatively distant power stations, the redevelopment of existing urban sites could have important benefits for electric utilities and the communities they serve. This could especially be the case of combined heat and electricity generation which is discussed later.

B. SITING OF NUCLEAR POWER PLANTS

In addition to the broader concerns about the utilisation of nuclear power, there remain many site specific concerns that have complicated the search for suitable locations for nuclear power plants.

The most important site specific concern is related to the risk of a major accident, induced by intrinsic causes or by external events (including sabotage) that would lead to the release of large amounts of radioactive materials. The problem of accident analysis and prevention has been treated extensively in many reports and publications. One of the best known is the Report WASH-1400 [Ref.9] (Rasmussen Report) which, despite the criticism that has been raised with respect to its findings from many quarters, remains one of the most authoritative references for nuclear power risk assessment.

The nuclear industry has a high safety record, the precautions taken against mishaps are stringent and the risks of catastrophic accidents extremely small. But as human nature is fallible, there is a possibility, however small, of a major accident. Siting essentially involves decisions which impose potential risks on people living within a certain radius of the site (0-10 km). The consequences of a catastrophic nuclear accident may be more severe than those from aircraft and several major industrial accidents, and may be of the same level of gravity as for low probability major catastrophies occurring in dangerous industrial facilities or due to natural events.

A basic criterion initially and still widely applied in all countries for reducing the potential impact associated with the risk of nuclear accidents is that of siting nuclear power plants in areas with very low population density and remote from sizable inhabited centres. With the increasing size of nuclear power programmes the available remote sites satisfying the other necessary requirements (low seismicity, availability of cooling water, transport networks, etc.) started to become scarce. At the same time significant progress in safety technology and assessment was being acquired with the installations. Therefore, several governments decided to license nuclear power station sites closer to urbanised areas. However, in some instances when siting was proposed within urban areas (e.g. Ravenswood, U.S.A.; Ludwigshafen, Federal Republic of Germany) licences were not granted (Ref. /10/).

The most recent development in siting policies is the agglomeration of two to six 1,000 MWe units in one site. Such an arrangement has some important technical and economic advantages. For example, a pair of nuclear reactors are built with common electricity generating installations, common cooling systems, and supporting and maintenance facilities. They need less operating staff than in the case of individual siting. Their operation may be adjusted in such a way that when one unit is being refuelled or under maintenance electricity production is not discontinued. Fewer construction workers are needed and the construction phase for such a complex is longer so that the labour force is permitted to settle for longer periods of time and the local authorities are able to depreciate the expenditure for the necessary infrastructure (see Chapter II b) over a much longer period. In addition to its undoubted economic and operational advantages and the public health improvement associated with a reduced number of nuclear sites and of population groups exposed to risk this siting policy may have also been a reaction to public opposition to nuclear power. In fact it is evident that regulatory procedures for siting four 1,000 MWe units in one site are much simpler than siting them in four sites. On the other hand, once the local inhabitants have accepted the first nuclear installation, much of their resistance is diluted when a second or third unit is added. Furthermore, once the initial impact to the influx of construction workers has been absorbed, the extension of their stay can bring only economic and social advantages to the area. Thus there is no doubt that this current policy of siting two to four units together has many advantages, provided that the site has the potential to absorb the physical environmental impacts.

This trend is illustrated in Table II-A2 which shows the ratio of operating or planned nuclear power stations to the number of sites as it was in 1970 and in 1985 or 1981 for some OECD countries.

Table II-A2

Country	Ratio in 1970	Ratio for 1981	Ratio for 1985
1. France	1.7	2.2	2.6
2. Germany	1.0	1.34	1.4
3. U.S.A.	1.0	1.5	1.9
4. Sweden	1.0	1.9	2.0
5. Japan	1.0	1.3	1.5

Calculated from the data of Table I-5.

The success of this policy and the shortage of new sites led to the idea of the "nuclear power park", i.e. the siting of a large number of 1,000 MWe LWRs - from a maximum of 40 units to a minimum of 10 - together with the necessary fuel fabrication, reprocessing and possibly radioactive waste disposal facilities. Such an arrangement will have a much reduced safeguard problem by the elimination of cross country transport of fuel and of waste, and some advantages in the employment of a construction force for 15-30 years in the same area. There are many arguments with respect to the economic advantages and drawbacks of a "nuclear park" (see, for example, reference [10]) but our purpose here is to examine its implications in siting.

Although it appears that the use of nuclear parks could reduce some of the problems relating to siting of nuclear power plants, not all of the effects are positive. More land is required per installed MW than if plants are located on dispersed sites, the very large amount of cooling water that is needed will be difficult to find, climatic effects may occur from the vast amounts of waste heat that must be dissipated and there are adverse amenity and land use impacts arising from the enormous concentration of transmission lines required to transmit the large blocks of power to consuming centres.

Underground siting has been studied (references [10] and [11] in many instances as a possible solution for siting nuclear plants close to urban areas. The main advantage of underground siting is that the overlying earth or sand or other partially treated material used for that purpose provides an additional reservoir for escaping steam and uncondensable gases, in case of an accident, thereby delaying any radioactive release in the atmosphere. Properly selected overburdens may act as an effective filter and an absorber for removing radioactive solid particles carried out as aerosols.

In the discussion at the IAEA-NEA Symposium in Vienna (December 1974) the cost of excavation was thought to be around 5 per cent of the capital cost of the plant for the United States, or that a 50m deep pit (soil) with backfilling will cost about $4 million. For the same size pit about $7 million should be spent in Germany. 100 m of cover will permit complete confinement of radioactivity except for gases and halogens and their escape will be at a much slower rate than would otherwise occur. However, costs of underground siting could increase the capital cost of nuclear generating plants as much as 10%. But, underground siting provides some economic benefits from lessened seismic loading, simpler foundation structures, shorter transmission lines - in case of siting nearer to load centres - and the possibility of district heating systems. Further studies will be needed to evaluate the economics of underground siting and the advantages it may provide in individual OECD countries.

Densely populated OECD countries like Belgium, Holland, Germany have been considering the creation of offshore islands for the siting of industrial facilities, particularly energy facilities. East Coast States of the United States have also been active in this field because of the need for deep water ports and for the siting of nuclear power plants. Thus a standardized offshore floating nuclear power plant is expected to be delivered by 1985 in the United States. The plant will be towed to its site - a breakwater constructed for the purpose.

An offshore nuclear plant opposite to a densely populated coast has many advantages, namely the availability of cooling water, the minimal use of land (for the transformer station and transmission lines) and the lack of adverse socio-economic impacts. There is no doubt that if safety considerations can be dealt with, offshore siting may provide an excellent solution for a number of OECD countries.

Comparison of Siting of Fossil Fuel and Nuclear Electric Generating Plants

Although siting of nuclear power generating stations creates some special problems, the decision as to whether to select a fossil fuel or nuclear plant at a given location involves consideration of other factors. As described in Chapter I nuclear power stations are more attractive than fossil fuelled power plants of the same capacity in the following ways:

- they are smaller in size, hence they use less land directly and have a smaller visual impact;
- they do not emit the large volumes of air pollutants that fossil fuel plants emit;
- they do not need the large transport systems (railways or ports and pipelines) for the transport of fuel or the removal of byproducts of combustion, hence they are devoid of the noise of railway and port traffic and of the pollution in terms of particulates and hydrocarbons in the air as well as of oil pollution of the sea; [transport of about 3,000 tons of coal or about 2,000 tons of oil per day for a 600 MWe station] ;
- routine releases of radioactivity, which used to be an issue, have been reduced to such an extent that they no longer seem to be a concern.

On the other hand, nuclear stations demand that the road system should allow for:

- the transport of the large, heavy nuclear containment equipment,
- the safe transport of fuel and radioactive waste,
- the rapid evacuation of the neighbouring population in case of an accident,
- larger volumes of cooling water because of the lower efficiencies of LWRs (32 per cent compared to 38-40 per cent for fossil fuel plants) and because they emit all their waste heat in the cooling water while in fossil fuel plants 10 per cent is emitted in the atmosphere. Thus, to'days nuclear power plants emit approximately 25 per cent more waste heat in water than conventional plants.

These substantial differences between fossil and nuclear power stations permit the choice of the technology to be made by considering both the particular characteristics of the site and its potential to absorb different kinds of environmental impacts. There is one recent case in the United States where a utility obtained a site and indicated to the regulatory authorities that it is ready to accept their advice on whether to build a nuclear or a coal fired power station.

C. COOLING SYSTEMS FOR POWER PLANTS

Another important constraint in the siting of electric power plants is the emission of "waste heat" and the associated requirements for cooling water. To put the problem in some perspective, a 1,000 MWe plant with a 38 per cent efficiency and 75 per cent load factor, would release about 7×10^{12} Kcals of heat annually, which is approximately the amount of energy required to heat 250,000 average houses in the United States.

There are four major options for providing the needed cooling: (a) once-through cooling; (b) cooling ponds; (c) wet cooling towers; and (d) dry cooling towers. Table II-A3 gives information on some of the important characteristics for these alternative systems as they affect siting, based on a 38 per cent efficient plant. These figures should be considered only representative estimates, as the characteristics can vary greatly depending on such things as climatological conditions, land costs, etc.

Once-through cooling systems have undoubtedly been the preferred choice of utilities in the past because of their low cost and the fact that they require no additional land for the power plant. The large water requirements for this cooling method dictate that the plant be located on the coast, along major rivers or on large lakes. However, because such areas have many other valuable uses (other industry, recreation, etc.) and because there have been growing concerns about the potential environmental effects of water temperature changes (especially where many plants are located along the same river) the number of sites available for once-through cooling have become more and more limited. It should be noted that although this option requires a large intake of water it "consumes" very little - the only consumption is that attributable to the increased evaporation caused by the higher temperature of the receiving water body, which is usually quite small.

In order to reduce the dependence on large amounts of water, one of the three other systems may be chosen. Cooling ponds are usually the next most economic and efficient system. However, they require substantial land areas which restrict siting possibilities in urban or near urban areas. The evaporation from the pond (both that which would normally occur and that induced by the heating) represents a significant amount of water consumption (although the exact amount is very sensitive to climatological conditions), which can restrict the use of this option in the very dry regions.

Wet cooling towers (natural or forced draft) require much smaller amounts of water than once-through cooling systems and have the advantage of needing less land compared to cooling ponds. For these reasons, they can give substantial flexibility in siting. On the other hand, wet cooling towers have raised public objections in several instances because of concerns with:

(i) the possible local climatological effects of the heat and water vapour emitted, which under some conditions could contribute to increased humidity and fogging; and,

(ii) the impact on visual amenity of the large tower structures and their associated plumes. For these reasons, wet cooling towers may still not be acceptable for urban or near urban siting of large power plants.

Table II-A3

TYPICAL CHARACTERISTICS FOR COOLING SYSTEMS
(based on 1000 MWe plant with 38 per cent efficiency)

Cooling System	Water (10^6 cubic metres/ year)		Land requirement (hectares)	Plant efficiency	Incremental cost of electricity above "once-through" *
	Intake	Consumed			
1. Once-through	1,140	small	0	38	-
2. Cooling ponds	34	21	420	38	6.0
3. Wet cooling towers	22	13.6	u	37.5	2.5
4. Dry cooling towers	.25	0	u	35	17.0

Source: Reference [6].

* In 1970 US$. Ref. "Engineering for Resolution of the Energy/ Environment Dilemma", NAE, Washington, D.C., 1972.

The last option in Table II-A3 is dry cooling towers. As indicated, this option eliminates completely the need for locating a power plant near a water source - only minimal amounts of make-up water for circulation are needed. Dry cooling towers would be even larger than the wet type, thus creating a slightly worse visual amenity problem but there would be no water vapour emitted to create a visible plume or fogging. For these reasons not only would they allow for siting in very dry regions but they might also make additional sites practical in urban or near-urban areas. This added flexibility in siting results in a loss of efficiency of around 8 per cent and an 8-10 per cent increase in electricity costs.

Another option, for which experimentation has been going on in some OECD countries, is the use of the waste heat of the water in agrotherm projects by passing it through pipes buried under agricultural land. Crop yields have been some 60% higher in experiments carried out in Germany while the cost of pipe laying was no greater than the construction of towers. This method appears to have merit both on economic grounds, in reducing the visual impact of cooling towers and, in the case of estuarine discharge in reducing the impacts of thermal pollution.

D. COMBINED HEAT AND ELECTRICITY PRODUCTION

In a conventional steam power plant the condensation temperature of steam is reduced as far as possible in order to obtain the highest possible conversion efficiency - typically 35-40 per cent. If the temperature of the cooling water is increased (thus increasing the condensation temperature of the steam), the electrical efficiency of the plant will drop. However, if the temperature is raised high enough, say to 100°C, then the "cooling" water can be used for such purposes as heating buildings (usually by district heating system) or as process heat or steam for industry. The loss of electricity due to the increased "cooling" water temperature is about 19-30 per cent, depending on the cooling water temperature, but the overall efficiency for this "combined" plant is about 85 per cent, compared to the 35-40 per cent for an electricity generation plant. The amount of heat produced in a large modern, combined district heating plant is about 1.5-1.7 times the electricity produced, while the ratio could increase to 2-3 for small district heating plants and small industrial plants supplying process steam. In summary then, combined heat and electricity generating plants offer a very attractive energy conservation option, since they allow for essentially full utilisation of the energy content in the fuel.

Because of its energy conserving character, combined heat and electricity production has found considerable application in some areas, primarily in Sweden and Finland. For example, in Finland one-fourth of the total electricity demand is generated in small and medium-sized combined plants within industry and in connection with district heating. During 1975 and 1976 these power sources corresponded to about half of the total thermal power generation in the whole country. Due to this, the average fuel consumption for thermal power generation in Finland was only 1800 Kcal/kWh, compared to a United States average of about 2620 Kcal/kWh.

However, despite its potential for utilising energy much more efficiently, combined heat and electricity production has not found significant levels of utilisation in most of the OECD countries. There are many complex obstacles to its development and a few examples will be described briefly.*

a) Economic aspects

It is not possible to give specific information here on the economic viability of combined heat and electricity production because it depends on many site specific factors, such as size and variation of the heat demand, fuel costs, interest rates, etc. However, the following simple example can illustrate the potential cost benefits*. Suppose that a combined plant has an 85 per cent

* More fully described in publ. Heat production and distribution, OECD 1978.

efficiency (based on the net heat content of the oil) and that it produces 1.7 kWh of hot water per kWh of electricity, so that the heat content that should be debited to the power production is 1.2 kWh-th per kWhe. In addition, assume that the fixed annual cost of a large, modern oil-fired power plant is about $70 per kWe (assuming 10 per cent interest, 20 years depreciation and a capital cost of $500 per kW). The corresponding, fixed annual cost for a combined plant of 200 MWe and 340 MWth (thermal) capacity is roughly $90 per kWe. The fixed annual cost for a large hot-water boiler is about $7 per kWth. Assuming an oil price of 4.18×10^{-3} cents per Kcal ($1.85 per million BTU), the minimum plant factor for an economically justified combined plant would be 15 per cent. While this example should only be considered as illustrative rather than typical or generally representative, it does indicate that combined heat and electricity generation can, in some cases at least, have some significant economic benefits.

For both industrial plants and district heating applications, the key economic variables are generally the size of the plant and the variability of the heat demand. A minor need for heat would mean of course a rather small back-pressure unit, implying higher relative costs for turbines, generators, piping, increased pressure of boilers, maintenance, labour, etc. If there are large and unforeseeable variations in the demand for heat, then the electric power supply availability is unreliable, with a corresponding reduction in its economic value, often two-thirds to a half of the value of reliable power. For district heating schemes, the cost of underground heat distribution is a very critical factor, and can vary substantially with local conditions. For example, it is estimated that the average cost of an underground network consisting of 500 mm diameter pipes would be about $500-1,000 per metre (1976 dollars) in Sweden, about $750-2,000 per metre in London, and about $10,000 per metre in downtown Paris. A rough estimate is that as a minimum a district heating system requires a heat demand density of 50,000 Kcal per square metre per year and a united area of 50.000 square metres in order to be feasible. Because of the costs and heat losses of long distance heat transport, it is generally believed that economic district heating applications will require relatively small plants (say 50-200 MWe) located close to the district heating network rather than large plants located outside the urban area.

b) Institutional aspects

However, even if standard engineering/economic calculations show combined plants to be economically attractive, a number of other factors of an institutional or organisational nature can block development. Two brief examples follow to illustrate these kinds of difficulties.

If the combined plant (either industrial or district heating) is owned by an enterprise other than the power company, then the electric power produced would have to be either sold to the electric company or used internally. Selling it to the power company would require a significant degree of co-operation and co-ordination in the scheduling of the other power plants owned by other enterprises or the electric utility. If the electric power is to be used internally by the combined plant owner, it must compete economically with purchased power. But the average price of purchased power will have an economic advantage, especially during periods of high inflation, because it will be based largely on the historical costs of existing plants and not on the true incremental cost of the additional power demand. Consequently, purchased power could erroneously appear to be cheaper than power from a new, combined plant, although the combined plant could be the best economic alternative from the overall system point of view. On the other hand, if the power company owned the combined plants, this could also create problems. For industrial operations, the production of steam is regarded as an integral part of the process, and the industrial enterprise would hesitate to have a foreign and autonomous organisation responsible for their supply of process steam.

Another difficulty affecting the development of combined heat and electricity production results from the differences in long-term financing of industrial versus electric power plants. In many OECD countries, power companies have free access to the capital market and are able to finance a major part of their capital expenditures through relatively low cost, long-term bonds. Government owned companies usually have an even more favourable position since they normally obtain the necessary funds from the Treasury. Normally a separate bond issue cannot be floated for an industrial combined plant because the investment might be too low to make a separate bond issue worthwhile and the combined plant cannot be considered separately from the remaining parts of the factory or industrial complex within which it is located. Loans would pay the same rate of interest as those for other industrial facilities which is definitely higher than that paid by bonds for conventional power plants. Perhaps more importantly, any such loan would inflict directly on the industry's remaining borrowing capacity, and most industries prefer to use their borrowing capacity for financing additional working capital and expanding their main activities and not for financing power producing equipment. Somewhat similar obstacles also exist for combined electricity-district heating plants belonging to municipality owned utilities. In their case it appears to be not so much a question of return on investment but more of a lack of capital resources.

c) Siting considerations

Having reviewed some of the advantages and obstacles to combined heat and electricity production, some of the implications for siting will be addressed, first for the case of district heating and then for industrial applications.

As indicated previously, for district heating applications small plants are usually located in the vicinity of the user due to the higher costs and heat losses that would be incurred in heat transmission with a large plant located outside the urban area. In order to discuss the changed nature of the siting problem caused by utilising combined plants, consider the following simple example which illustrates the two extreme cases. Suppose Option 1 is to build a large 600 MWe power plant somewhere away from the urban area and eight 150 MWth boilers for district heating within the city; and Option 2 is to build eight combined plants in the city each with 75 MWe and 150 MWth capacity (1). Since the energy efficiency for Option 2 is substantially higher, a reduction of 25 per cent in amount of primary energy consumed (2), the total pollution levels would be reduced proportionately and the waste heat would be non-existent (except during those periods when the combined plants run as straight condensing plants).

Another factor in favour of the combined plant system (Option 2) is that it would eliminate the need for a single, very large plant (600 MWe). Utilising combined heat and electricity plants is one example of the so-called "small is beautiful" concept, which is in contrast to most current utility company inclinations toward huge power plant schemes - 500 to 1000 MWe. As pointed out in Chapter I, much of the opposition to MEFs can be attributed to the magnitude and concentration of the adverse environmental and socio-economic impacts that a large energy facility can cause. With small plants not only would many more sites be technically feasible (e.g. they could be located along small rivers) but they are not as likely to arouse public opposition since the magnitude of the impacts would be smaller. (The socio-economic impacts are discussed at some length in part II of this Chapter.) In addition to these factors, it has been argued that there are economic disbenefits associated with very large plants which are often not foreseen in utility company planning.

---

1) It will be assumed that the combined plants are flexible (i.e. can run as straight condensing plants) so that the electricity availability factor is essentially the same in both cases.

2) Assuming an overall capacity factor for the entire system in both cases of 50 per cent, Option 1 would require approximately 556,000 tonnes of oil annually in the heating plants and 600,000 tonnes at the power plant; Option 2 would require 906,000 tonnes of oil annually, assuming an average efficiency of 75 per cent to reflect some operation as straight condensing plants.

For example, the required lead times for small plants are much shorter, thus, reducing substantially the interest and escalation costs normally incurred with large plants, and also reducing the risks due to incorrect demand forecasts or the risks due to construction delays for very large plants. Smaller, dispersed plants could save large amounts on the overheads associated with large plant systems - transmission lines, transformers, management and engineers, etc.; and small plants give greater flexibility in scheduling and thus may allow for smaller levels of reserve capacity. Whether these advantages of small plants generally outweigh the economies of scale of large units is dependent on a number of conditions specific to each application. Nevertheless, where combined plants can compare favourably with conventional systems in the usual economic and siting terms advantages associated with "smallness" given even more weight in their favour.

Other factors that must be considered in evaluating the siting implications of combined heat and electricity production relate to the fact that they will be located within the urban area. This has the advantage of reducing the number of high-voltage transmission lines coming into the city, but, on the other hand, locating small plants in major population centres brings the associated environmental problems closer to people. Both Options 1 and 2 (page 70) require the same number of sites within the city, but the combined plants will be slightly larger and will burn about 62 per cent more fuel in the city for the situation given in the above illustrative example. Probably the most serious problem to contend with is air pollution and strict limits on emissions will probably be required in both cases. From the discussion in the previous section on air pollution control alternatives, by far the best option for these small combined plants from a siting point of view, is the use of "clean" fuels and/or the use of tall stacks. Certainly, the "throw-away" systems will require additional land for process equipment, waste water treatment, etc.

The transportation of fuel to local plants may also become a significant factor in the siting decision, not only because of the direct costs but also because of its indirect effects. For example, the transportation of fuel could disturb other public transportation, making it necessary to carefully evaluate alternative transportation possibilities (water, railway and highways) in the selection of the site. But again, the requirement for fuel transportation into the urban area will not be a new factor for combined plants, but will only be increased proportionately to the increased fuel consumption within the city.

In examining industrial application, if an industrial plant would normally generate its own electricity and heat needs separately, then the effects on siting of combined generation would be strictly positive - fewer emissions per unit of production due to the higher energy utilisation efficiencies. On the other hand, if the industrial plant would normally purchase power produced externally, then the combined case would face many of the problems discussed above for combined plants for district heating, including an increase in energy consumption and of the environmental problems for the plant. However, since such plants are normally located in industrial zones anyway, site acquisition should not be significantly more difficult.

In summary, combined heat and electricity production reduces environmental pollution levels due to the improvements in energy utilisation efficiencies, and for this reason it should reduce siting difficulties for energy facilities. However, because they can account for increased pollution emissions in urban areas and because optimum siting arrangements are dependent on a variety of factors specific to each application, each case must be examined on its own merits to determine the potential advantages. Nevertheless, experience to date in Finland and Sweden has shown that reducing the environmental impacts of local combined plants is not a decisive cost factor and no significant siting obstacles have been encountered.

E. CONCLUSIONS - PART I

Reducing the adverse physical impacts caused by an MEF should give increased flexibility in siting, either by expanding the number of sites which are technically available (e.g. by reducing water requirements) or by reducing public objections on environmental grounds. However, any attempt to control adverse impacts usually incurs a penalty in efficiency terms; and many times reducing one impact can create more difficult problems in another area (e.g. reducing water requirements can increase land use). In addition, "optimum" solutions are very dependent on the specific situation being considered - regulations in effect, existing ambient air quality, water availability, etc. For these reasons, it is not possible to offer general solutions as to the "optimum" technology for any type of MEF.

i) The siting decision should consider explicitly the alternatives available and should weigh them in terms of local conditions, regulations, economic costs, etc. In other words, the siting process should not be viewed as simply one of trying to decide where to locate a given plant but rather should be framed in terms of what combination of technology and location would best serve the public interest.

ii) There is little doubt that many of the public's concerns which have constrained the siting of MEFs are attributable to the magnitude of impacts created by the large scale of such facilities. For this reason, the potential advantages of smaller energy facilities from environmental, economic and efficiency standpoints should not be overlooked. This may include increased R & D efforts for technologies that can best be utilised on a small scale.

iii) Combined heat and electricity production has definite advantages in energy efficiency and environmental terms compared to alternatives. These factors, in conjunction with its "smallness", mean that it should also reduce siting difficulties. In addition, experience has shown that it is economically attractive in some instances. Despite these facts, there are many constraints due to organisational or institutional factors which have blocked its utilisation. Therefore, countries should give increased attention to combined heat and electricity production and should work to remove institutional or organisational barriers that prevent its use where it would otherwise be economically feasible.

iv) Among technical methods, the use of clean fossil fuel provides the greater flexibility for siting to utilities, energy consuming industrial plants, combined heat and electricity production plants, and other energy consumers. Given the current developments in energy and siting policies, Member countries should evaluate the use of clean fuels in technical, economic, efficiency and environmental terms.

Part II

ANTICIPATING THE ADVERSE SOCIAL AND ECONOMIC IMPACTS

The social and economic impacts of MEFs on the local community are similar whether the MEF is a nuclear power plant, a refinery or a construction site for offshore platforms. The probable social and economic impacts are the following: increase of population; change in the composition of the local community with resulting changes in culture and life styles; changes in the labour market; pressure on housing, schools, transport systems, public transport, social services, hospitals, recreational facilities and entertainment; increased local business; increased local taxation income and further industrialisation. The similarities derive from the following:

a) all these facilities require a long construction period (four-ten years), a big construction site, and a large, skilled construction force;

b) once the construction has been completed, a much smaller but more specialised, workforce takes over;

c) local income increases substantially through salaries, increased business and local taxes;

d) considerable pressure is put on a local authority to provide housing, school and recreation facilities, roads, public transport and social services, particularly during the construction period.

The difference between physical impacts of an MEF on the one hand, social and economic impacts on the other is that the latter can be overall beneficial for the local and regional communities. Indeed, the beneficial economic and social effects - local employment, local taxation and increased local business - have been the causes for the eagerness of local authorities to accommodate MEFs. However, the extent to which local communities may benefit from the siting of an MEF depends on the extent of co-operation between the developer, the local and regional authorities and the public when planning the different stages of the development. If this fails then the overall effect may prove detrimental to the community.

A factor which may influence events to the extent that benefits for one locality may prove disbenefits for another is the extent of industrialisation of the community. An industrial area may absorb an MEF with fewer adverse social or economic impacts since housing, schools, public transport, roads and other infrastructure are there, as well as labour, skills, etc. Even in times of full employment, when workers have to move in from other areas and some pressure on housing may develop, there will not be a substantial social change in the composition of the community. Hence, developers often prefer to site MEFs next to industrial and urban areas.

This is not, however, always possible. The nature of the technology, e.g. nuclear power plants, LNG terminals, etc., the excessive concentration of industry in some areas, the need for economic development of other areas, and the geographic location of energy resources (North Sea oil, coal in the North-Western United States, hydro-electric power, etc.) call for the siting of MEFs in areas where population may be scarce, and industry non-existent. In such cases, unless careful planning is carried out and local authorities are provided with the financial means to build the required infrastructure and services, through taxation, subsidies or by co-operation with the developer, the community may be harmed socially and economically.

A. EMPLOYMENT DURING CONSTRUCTION AND OPERATION OF AN MEF

In considering the social implications, and to assess the impact of the project on the existing community, it is important to know as accurately as possible the numbers of employees needed at the various stages of development, the different skills involved, the percentage of employees to be recruited locally, proportion of male to female, number of key workers, etc. Table II-B1 gives the total, peak and annual employment for the construction period for a number of MEFs in some OECD countries, the total amount of salaries paid to these workers and the part of that income spent locally. It also presents figures on the capital investment of these MEFs, the employment and salaries paid during operation. Although there are substantial differences among MEFs and among countries, it is evident that:

i) a big construction workforce is needed;
ii) the MEFs are highly capital intensive;
iii) about 75 per cent of the income of construction workers is spent locally;
iv) a much smaller working force is needed to operate the MEF.

The effect of such an influx of workers in a remote community can be easily appreciated.

The benefit of employment will be greatest if there are local unemployed skills at the beginning of the project. If not, the developer will recruit labour from local industry, to which higher wages are offered. Local employers may be unable to compete and local industry (agriculture, construction, etc.) may be adversely affected. The more basic local services may also be harmed in that the labour force employed on road maintenance, railways, electricity, water and sewerage installations require a minimum of training to join the developer's workforce.

As can be seen in Table II-B1, an established feature of most MEFs is that the construction workforce will considerably exceed the permanent workforce. The latter's composition is also different from the former's in that its range extends from plant managers and company executives to engineers, technicians, skilled and unskilled labour. If local industry was forced to close due to loss of labour, when the construction force is disbanded many of the previous jobs may no longer be available. Thus the completion of large-scale construction schemes has often been followed by local unemployment. This is due to the difficulty of finding new employment at the end of the construction period, unless there is a new large-scale project in the area. The situation becomes worse in a period of economic depression since it is difficult to create new jobs for local workers, and migrant workers tend to stay in the area, adding to the number of unemployed. Thus the long-term effect of such big construction sites may be that they help to prevent the growth of employment in more stable industries.

There seems to be two possible ways in dealing with that problem:

Table II-B1

EMPLOYMENT AND ECONOMIC IMPACT OF MEFs

	TYPE OF FACILITY	CONSTRUCTION PHASE						OPERATION PHASE	
		Employment			Finance				
		Total number of workers (*)	Number of workers at peak	Annual average employment	Invest-ment (**)	Total salaries (**)	Portion spent locally	Employ-ment (*)	Salaries (**)
1	Refinery (United Kingdom) 200,000 b/d	3,500	1,500	1,000	300			350-450	
2	Nuclear plant (France) (~ 2,000 MWe)	6,000	1,250	800	500	20	15	210	3
3	Nuclear plant (United States) (~ 4,400 MWe)	15,000	3,450	2,000	2,200	250	190	150	4,5
4	Nuclear plant (United States) (~ 2,400 MWe)	16,500	3,100	1,840	1,900	264	200	100	3,9
5	Nuclear plant (United Kingdom) (~ 1,650 MWe)	12,000	2,000	1,350		300-400		600	
6	Nuclear plant (United Kingdom) (***) (~ 1,180 MWe)	10,000	2,400	1,400	300	96		580	
7	Scanitro (United Kingdom) Ammonia plant (1,200 t/d)	400-600	300		100			100-120	

(*) In man/years.

(**) In million US$ (1972).

(***) Including a pumped storage station.

a) Construction workers, because of the nature of the industry, tend to move quite easily from one building site to another. It may, therefore, be preferable for the local economy, whenever the local unemployment level is low, that the developer does not take advantage of the higher salaries he is paying and of the bonuses he distributes to expedite the work to lure labour and skills from local industry, but rather that he "imports" skills from other areas.

b) Governments, whenever MEFs are being installed away from established urban and industrial centres, consider the possibility that other industrial installations be sited in the same area, so that the construction labour remains for a longer period and the infrastructure provided locally does not have to be dismantled once the MEF has been completed.

B. HOUSING, SCHOOLS, PUBLIC TRANSPORT, ROADS, SERVICES AND OTHER INFRASTRUCTURE

The problems associated with the population shifts of the construction phase of MEFs have attracted not only the attention of local communities, but also of regional and even national authorities. The Norwegian Government, for example, has been looking very carefully at the labour needs in the case of developing substantial platform construction and petrochemical units along the North Sea coast which may lead to people, particularly the young, leaving their villages and agriculture for the cities and industry. Similar concerns have been expressed in Scotland, despite the chronic unemployment and emigration, for a labour force of the size indicated in Table II-B1, the pressure on the local infrastructure of non-urban areas is considerable (1). To lessen difficulties caused by an influx of construction workers some developers recruit only unattached men, thereby cutting down on the requirements for housing and education, but there is nothing to prevent a man bringing his family to the area and housing them in caravans, mobile homes, or other temporary accommodations, which in turn can cause problems.

Housing may become the most difficult problem that local authorities have to face. Again, remote, less populated areas may suffer most. Cities and densely populated areas can accommodate the construction force much more easily. In Belgium, for example, most of the construction force for the nuclear power plants at Tihange and Doel was commuting, or staying in hotels and returning to their homes for the weekend. In the United Kingdom, CEGB seldom has to erect working camps. In most cases construction workers find accommodation

1) For a detailed description of social and economic impacts of MEFs see the OECD Document referred to in page 5.

in nearby villages and small towns and commute to the building site. On the contrary, as shown in Table II-B2, for the nuclear plants at Fessenheim and Bugey in France, only 21 per cent and 13 per cent of the workers respectively were accommodated in local houses and apartments, while the bulk, 58 per cent and 73 per cent respectively, were accommodated by the developer in hotels and company buildings and the rest dwelled in caravans.

The adequacy of schools is a very important matter for both construction and permanent staff. Planning for schooling needs is not an easy task when the fluctuations of the workforce, described in Table II-B3, are taken into account. To this must be added the mobility of construction workers which may cause further complications. An enquiry of the schooling needs at Fessenheim and Bugey gave the figures of Table II-B4 for different levels of schooling.

Table II-B2

TYPES OF HOUSING AT THE FESSENHEIM AND BUGEY NUCLEAR PLANT CONSTRUCTION SITES

Type of Housing	Fessenheim Site 2 PWR x 925 MWe		Bugey Site 4 PWR x 925 MWe	
	No. of units	%	No. of units	%
1. Houses and apartments	169	21	197	13
2. Developers' buildings by the site	77	9	609	41
3. Furnished rooms and hotels	421	51	497	33
4. Caravans (trailers)	163	20	194	13
TOTAL	830	100	1,497	100

Table II-B3

NUMBERS OF CONSTRUCTION EMPLOYEES AT FESSENHEIM AND BUGEY

Site	1971	1972	1973	1974	1975	1976
1. Fessenheim	100	312	604	812	861	
2. Bugey		91	471	1,224	2,054	2,600

Table II-B4

NUMBER OF CHILDREN AND AGE BRACKETS AT FESSENHEIM AND BUGEY CONSTRUCTION SITES (*)

	Total	Age brackets			
		Babies	Nursery	Primary school	Secondary school
Construction force					
Fessenheim	679	103	189	278	109
Bugey	468	85	114	216	53
Personnel EDF at each site	87-119	18-21	22-30	39-53	12-15

By the end of the fourth year of construction, fluctuations in the number of children of schooling age become less pronounced until the departure of the construction force the arrival of operating staff introduces another upheaval. Although there is not yet experience with the operating stage, EDF has predicted the number of school-age children of operational staff for a 2 x 1,000 MWe and a 4 x 1,000 MWe nuclear power plants, as shown in Table II-B5. These figures are for the first year of operation.

Table II-B5

SCHOOL AGE CHILDREN OF OPERATING STAFF FOR NUCLEAR PLANTS; ESTIMATED BY EDF

	Total	Babies	Nursery	Primary	Secondary	Higher
2 x 1,000 MWe	300	66	69	90	66	9
4 x 1,000 MWe	525	115	121	157	115	16

An MEF also puts substantial burden on the transport systems of the area, and in many instances requires new roads, railway tracks,

(*) The reference for EDF-data, "Impact Démographique, Economique et Social de l'implantation d'une centrale nucléaire, EDF, January 1976", does not indicate the year for which these figures have been collected. It indicates however that for Fessenheim there were 528 families, and for Bugey 426 families. Thus from Table II-B3 it may be derived that for Fessenheim the year must be either 1974 or 1975 (given the fact that 50-60 per cent of the workers bring their families) but for Bugey the evaluation must have been made somewhere between 1973 and 1974.

ports, etc. During the construction period the transport of heavy equipment, of thousands of tons of building materials, and the commuting of large numbers of construction workers may put a heavy pressure on the existing road network and on local public transport. During operation the transport of fuel (oil, coal, etc.) of products (gas, refinery products, etc.) of used fuel (nuclear power stations), etc., can sustain the pressure on transport systems.

The influx of a large number of workers and their families puts considerable pressure on social services, hospitals, police, entertainment facilities, etc., particularly when the community is a small one. There were instances e.g. in Scotland there this has produced considerable resentment on the part of the local population which had to accommodate newcomers whose numbers exceeded that of the inhabitants. The availability of entertainment is important because in construction projects many of the workers are young single men who earn substantial wages. Whenever possible, the working force may be dispersed to a number of neighbouring towns or villages, but in any case local and regional authorities should plan to provide adequate facilities to avoid upsetting the local communities.

C. ECONOMIC IMPACTS DURING CONSTRUCTION AND OPERATION

a) Costs of infrastructure

Local authorities have to provide in most OECD countries the infrastructure needed for the construction and operation of an MEF. Although, as will be described later, the local taxes paid by the developer are substantial and in many instances raise the local authority's income manyfold, traditionally they are paid only when the MEF goes on stream. Hence the local authority has to borrow the money or be subsidised by the regional or national government. There are instances, e.g. in Scotland, where the local authority, to reduce local unemployment, offered to undertake the financial burden of providing the infrastructure. In other instances, e.g. the development of the Fos-sur-Mer industrial zone, considerable difficulties were met in housing and transport from the lack of infrastructure. It does, therefore, seem justified that ways are found to provide local authorities in a timely manner with the information and financial resources necessary for providing the infrastructure.

Several countries have developed procedures for providing the required finances for infrastructure in the time frame needed. In Japan, for example, the electric companies are required to pay to the Government 40 cent per 1,000 kWh they produced and the Central Government uses this tax revenue for providing grants to local governments which need to finance infrastructure. The amount paid is $2 per year per kW installed, which is split in half to the municipality that houses the power station and to the neighbouring municipalities. The grant is given for the period between the start of construction and the start of plant operation, and in this respect this may be regarded as a sort of advance payment. For a 1,000 MWe power plant this grant amounts to $ 2 million per year.

In France legislation was passed recently (1976) regarding large construction sites which deals with the financing of infrastructure in the following way:

The infrastructure is divided into two categories: (a) infrastructure specific to the construction site less to the community once the project is completed. In this case the expenditure needed is made part of the total investment. And (b) infrastructure which will be used by the community and which is built in advance because of the needs of the construction site. This is financed by government subsidies or by the local authority. Whenever the local authority has to resort to borrowing the money, the developer is asked to pay in advance the annual taxes he will be paying once the MEF goes on stream.

In Canada provincial corporations like Ontario Hydro are not subject to taxation for municipal or school purposes. However, it is required to pay grants in lieu of taxes to local jurisdictions based on a set method of evaluating the plant and with the limitation that they should not exceed 50 per cent of the total other taxes of the municipality. With respect to the 4 x 1,000 MWe Bruce nuclear power development, Ontario Hydro decided in 1970 to pay during construction an extra $100,000 per year for ten years in addition to the grants in lieu. Further agreements with the local authorities led the company to agree to a payment of $500,000 to continue for the life of the project. Another $120,000 was paid for road damages, and lastly $5 million will be paid for building a new road.

A recent study (Reference [12]) indicates that the creation of "nuclear parks" will pose considerable problems in the distribution of costs and benefits amongst local, regional and central (federal) government. Furthermore, the current system of distributing costs and benefits among local residents and users at large through taxation will collapse as benefits up to $70,000 per person per year may accrue for local residents. Even with a 4,000 MWe site, in some areas, these benefits have risen to $8,000 per person per year. Thus it seems that some experience from smaller unit sites should be acquired with respect to their environmental and socio-economic impacts before decisions are taken to create much larger concentrations of generating capacity.

b) <u>Local business</u>

The economic impacts on the local community resulting from the influx of the construction and, later, the operation staff is considerable. Some evaluations of these impacts have been carried out in OECD countries and give an indication of the size of effects that may be expected. Several examples follow.

Table II-B6 shows the effect of a nuclear power plant on the local economy of Gwynedd in Wales in terms of increased income and increased business activity during the construction and operation phases. Table II-B7 shows the distribution of the income from salaries and other expenditure of EDF in the Fessenheim and Bugey sites.

In both cases the annual incomes from salaries during the construction phase amount to between $14 and $16 million with a considerable amount of that income spent in housing and food. The EIS on the Vogtle (Georgia, United States) nuclear power station (4 x 1,000 MWe) states that a workforce of 15,000 per year on an average income of $16,000 will spend about 75 per cent of its total income of $250 million, i.e. about $180 million, in the locality. There can therefore be little doubt as to the financial advantages for local business from an MEF.

c) Local tax income

Local tax income from MEFs is an important contribution to the finances of a local authority. This is, of course, also the case with other industrial facilities, but the case of MEFs tends to become exacerbated due to their high capital intensity, low employment, and the fact that in many instances they are sited in sparsely populated areas.

Although local tax rates are calculated in different OECD countries in a different way and tend to vary among local authorities of the same country, the total taxes paid by an MEF are in most cases substantial. For most OECD countries local tax rates are based on property values, turnover of an enterprise, salaries paid, or by combination of these factors. Local authorities may choose to use that income in improving the services they provide to the community, or to reduce the local tax rates thus providing a direct economic benefit to the inhabitants, or both.

Local tax systems have been changing in many OECD countries, and these have affected MEFs. One reason for the changes is that large sums were being paid to small communities, while the social and economic effects were felt in a much wider area. For example, in France the old system of the "patents" which was based on the application of locally determined tax rates on a nationally established taxable basis was changed when it was found that very small communities became recipients of very large sums which could not be used fruitfully, while neighbouring communities had to sustain most of the social and economic effects - particularly during the construction period. The new system (Taxe professionnelle) is based on

Table II-B6

THE EXTRA INCOMES AND INCREASED BUSINESS ACTIVITY EXPECTED TO BE GENERATED IN SECTORS OF THE GWYNEDD, WALES (UNITED KINGDOM) ECONOMY DURING, YEAR THREE OF THE CONSTRUCTION PHASE YEAR OF MAXIMUM EMPLOYMENT, AND IN YEAR TEN OF THE OPERATION PHASE (in United States $) - /The MEF is a SGHWR 1,000 MWe nuclear reactor/.

Sector	Construction phase		Operation phase	
	Increase in incomes	Increase in business activity	Increase in incomes	Increase in business activity
1. Primary industries (agriculture, eng. etc.)	187,200	495,600	47,800	125,200
2. Manufacturing, gas, electr., water	235,600	1,032,200	64,000	279,800
3. Transport, post, tele., banking, education	562,200	1,625,600	165,200	282,400
4. Hotels, catering, food	1,751,600	9,764,000	469,000	2,594,000
5. Other services	730,000	1,724,000	225,800	528,200
Sub total	3,466,600		971,800	
Initial income injection	15,600,000		4,186,200	
Less unemployment benefits	2,433,600			
TOTAL	16,633,000	14,641,800	5,158,000	4,009,600

Reference: The impact of a power station on Gwynedd (Mimeo, 1976) by C. Chadwick, County Planning Officer, Study supported by CEGB and the County Council.

the taxable value of an installation, the salaries paid to employees and the local tax rates. There is, however, a limit of $1,000 per person per year that should not be exceed. In the latter case the extra tax income is distributed among neighbouring communities, regional authorities and neighbouring regions according to the extent of their support to the development.

Exact figures as to sums paid to local authorities by different MEFs are difficult to obtain (1). It is expected that three nuclear

1) The examples given in this paragraph and mid page 87 are from the OECD Document referred to in page 5.

Table II-B7

DISTRIBUTION OF ANNUAL SALARIES AND TOTAL EXPENDITURE DURING THE CONSTRUCTION PHASE OF NUCLEAR POWER PLANTS IN FRANCE

Sector	On salaries	
	%	$Millions
1. Food	21	4.2
2. Hotel, restaurants	18	3.6
3. Housing	9	1.8
4. Transport, tele.	7.5	1.5
5. Clothing	5.5	1.1
6. Other services	7.0	1.4
Subtotal	68	13.6
Taxes	2	0.4
Insurance	3.5	0.7
Subtotal	73.5	14.7
Savings or spent elsewhere	26.5	5.3
TOTAL	100.0	20.0

power stations in operation in France will pay the following sums according to the new law:

Chinon (690 MWe) $1.76 million
St. Laurent (1,000 MWe) $3.8 million
Le Bugey (540 MWe) $1.34 million

Of the above sums about 60 per cent will be paid to local communities and 40 per cent to be distributed by the regional authorities. Thus for the community of St. Laurent which numbers 1,750 inhabitants, the sums should have been about $1,300 per person per year. The $300 per person above the maximum $1,000 will be distributed as explained in the previous paragraph.

In the United Kingdom the total tax paid for electricity generation and distribution amounts to about $297 million. Half of that ($150 million) is paid by the regional distribution systems. The other half is paid by the local authorities for the distribution network on the basis of their population, size, industrial production, etc. The rest, 75 per cent ($120 million), is paid to local authorities for electricity generation on the basis of electricity generated within their area and local tax rates. Although the latter vary among local authorities $1.7 per kW generated is considered an average figure for CEGB. On that basis a 1,000 MWe power station will pay to the local authority about $1,700,000 per year.

In Japan, the major tax paid by a nuclear power station to the Central Government is the Fixed Assets Tax which is calculated as 1.4-2.1 per cent on the investment. For a 1,000 MWe station this investment is about $530 million, hence the tax is $7.42-11.13 million. The community which houses the station is given about 25 per cent of that sum, i.e. $1.86 million, the rest being distributed through Central Government. There are, however, other taxes such as the Business Office Tax, the Real Property Acquisition Tax and the Special Landholding Tax.

In the United States taxes paid vary widely among States and among counties within a State. According to the EIS prepared for a number of nuclear power stations the following figures are reported:

i) For the Perry (Cleveland) 2 x 1,200 MWe nuclear power station (Investment $1,234 million) the expected annual taxes will amount to:
$34 million for the school system;
$8.7 million for the local and county authorities.

ii) For the Vogtle (Georgia) 4 x 1,100 MWe nuclear power station (Investment $2,200 million) the taxes will be calculated as follows:
$10.530 million for the school system;
$4.455 million for the county system;
$.202 million for the state system.

In Belgium, annual land use and tax income paid by a utility operating a 900 MWe nuclear power plant is expected to be:
$300,000 to $510,000 for the provincial authorities;
$730,000 to $1,270,000 for the local authorities.

Whatever the system on which local taxes are based, the high investment and the area needed for an MEF makes the tax revenue substantial, and hence the MEF is often welcomed by local authorities. MEFs, however, have to compete with other industrial installations which may pay similar or even higher local taxes and may provide other advantages, e.g. higher and more permanent employment to the local community. For example, in a case in the United Kingdom (the Connah's Quay power station) the local authority had agreed to siting a nuclear power plant but changed its decision when the government indicated its interest in expanding housing projects in that area to allow the settlement of industry employees of neighbouring communities. For this particular local authority, enlarging the constituency and increasing local business was more important than housing a nuclear power plant and acquiring the associated tax revenue. There can be little doubt that this system of competition among industrial installations in getting sites is a healthy system which provides choice for local authorities and reduces the possibility of creating environmental "black spots", as they can decide to accommodate both polluting and non-polluting industries.

D. SOCIO-ECONOMIC EFFECTS: A MATTER FOR PLANNING AND COMPROMISE

It is possible that the installation of an MEF and the promise it brings of further industrial development in the area (refineries leading to petro-chemical manufacturers, power plants attracting energy intensive industry like aluminium smelters, deep sea ports needing tank farms, pipeline terminals, etc.) is resisted by local authorities and communities whose wealth derives traditionally from other industries like agriculture, fishing, tourism, etc. In such cases, even when the balance and influence of local political groups may be affected, there are grounds for bargain and compromise between developer and local authorities. Here, regional authorities may be best placed to evaluate the pros and cons of the development in a wider than local context and play the role of arbitrator. There are many examples in OECD countries when compromise was reached and local interests were safeguarded while MEFs were installed. For example, the Scottish fishing industry has been provided with new harbour facilities to replace those utilised by the offshore oil industry. Another example is the offer made in Britanny (France) by the Government to help safeguard substantial parts of the coast from any development to counterbalance the use of other parts of the coast for siting nuclear power plants. What is important in this context is that regulatory procedures be structered in such a way as to allow for such bargain and compromise to take place.

From the above it is evident that an MEF enhances, in most cases, the economic development of a community. Employment opportunities and the increase in local business is the most important factor. Increased local tax income may provide to the community the means to acquire or build recreational facilities, parks, etc. and in general to improve the physical and social environment. Furthermore, the operation of an MEF may have other effects on a community. The installation in the area of 300-500 new households coming from different social backgrounds and belonging to different income brackets may activate dormant communities, particularly when the newcomers are encouraged by the developer to participate in the activities of the community. New community facilities may take shape, shops may start to sell a wider variety of goods, transport services or better roads responding to greater demand may be built, better services may become available when a middle class, which may have been lacking before, is introduced in an area and the average age of the permanent workforce may be appreciably below that of the community, thus adding to its vitality.

However, all these possible advantages for a local community that may result from the siting of an MEF may fail to materialise and even adverse impacts be created instead, if the developer and planning authorities have not ensured:

a) That the social and economic structure of the community and the local public's concerns have been taken into account, and that there is no widespread outright opposition to the project;

b) That care is taken and the means have been provided to create the infrastructure that is needed to avoid the severe social and economic impacts of the construction phase and to allow for a smooth transition to the operation phase.

E. CONCLUSIONS - PART II

The economic benefits from an MEF on a local community have been recognised for a very long time and have always provided the arguments to local authorities for accepting such facilities. The advantages provided for local employment, local business and for the improvement of the service sector through higher local tax income are very real and have become more important because of the large size of MEFs. But for the same reason, such developments should be planned very carefully to anticipate the impacts, particularly of the construction phase, on local society and local economy:

(i) Social and economic advantages from the construction and operation of MEFs are increased if the site is chosen within an industrial area with some unemployment.

(ii) If the site chosen is not within an existing industrial area, consideration may be given that the area around the site has the potential of becoming an area of industrial development. In this way, the construction force may have to settle in a permanent way and be eventually absorbed in the operation staff of industrial installations.

(iii) A part of the local taxation which is to be paid by the developer when the MEF goes into operation may be paid in advance, either through government loans or by the developer so that the local authorities are able to finance the accessory infrastructure in time to avoid adverse social impacts.

(iv) The local public identifies, together with the local authorities and the developer, the probable social and economic impacts to be assessed, and is given the opportunity to discuss and comment on the findings of the assessment.

(v) The adverse social and economic effects that may result during the transition from the construction to the operation phase should be studied in advance and consideration be given to the measures which will be necessary to reduce those impacts.

Chapter III

SITING POLICIES AND PROCEDURES

A. THE NEED FOR NEW SITING POLICIES AND PROCEDURES

A number of important factors relating to the siting of MEFs have changed markedly over the past 10 to 15 years. These include such physical factors as (a) a very large increase in the number of facilities required to meet the energy needs of a greatly expanded level of GNP in the major industrialised countries, (b) the continued growth in the size of the individual facilities in order to obtain economies of scale, and (c) the introduction on a greatly expanded scale of new technologies, such as nuclear generating plants, with entirely new and highly specialised siting requirements. There have also been major changes in the political and social factors affecting MEF siting. The most important of these are: (a) the desire of Member countries to achieve a greater degree of security in energy supplies, (b) the emergence of strong environmental and consumer movements designed to assure that not only are interests taken into account in the decision-making process but also that means be devised for meaningful citizen participation in the decisions, and (c) a cultural change in which an increased emphasis is being placed in achieving an improved quality of life rather than merely striving for a larger per capita income.

These changes - physical, political, social and economic - in the nature of energy facilities and the economic and cultural structure of OECD countries have required that major changes be made in the participants, the mechanisms and the regulations that are involved in the siting of MEFs.

Although significant improvements are taking place in the siting process in many Member countries that are responsive to the changed conditions discussed above, there still remains room for further improvement.

This Chapter will discuss the role of the various regulatory principles that are being used, and how these have been altered over time to accommodate the new factors that have become important in the siting process. Examples are given of developments in various OECD countries aimed at changing procedures to accommodate the siting process both to the new participants and to the drastically changed siting problems and objectives. Finally, a new framework is proposed for siting to further improve the efficiency and effectiveness of the process - a framework designed to be flexible enough to respond to new physical, political and social changes as they occur and to the constitutional constraints of Member countries.

B. THE PARTICIPANTS IN THE SITING PROCESS

Until recently the siting of MEFs with the exception of nuclear power plants has been to a large extent the responsibility of the developers and of local authorities. The former being responsible to either government or to their stockholders and to the public they serve to meet the demand for their products and make a profit; the latter being responsible for the use of land. National and regional governments have had a less direct role in the process but this has tended to become more important as the difficulties of siting increased. Moreover, as the size of the installations have become much larger, their adverse effects - environmental and social - have greatly increased. The impacts are felt beyond the local level and thus a larger number of citizens have sought to participate in the decision-making process.

The developers

Among the increasing number of participants in the siting process, the developers carry, perhaps, most of the financial risk.

The developers are :

(i) large public quasi monopolistic companies like CEGB in the United Kingdom, EDF in France, the Public Power Corp. in Greece, etc., whose total employment is counted in thousands and which have strong ties with the Administration;

(ii) privately owned utilities, like Pacific Gas and Electricity in California which operate in most cases on a regional basis;

(iii) multinational firms, particularly those of the oil industry, whose strategies are more or less global;

(iv) an assortment of national companies, consortia, and other groups of companies which are created for a particular project.

Their activities with respect to siting are the following:

(i) to evaluate the demand for their product, to identify the need for a new facility and determine its size;

(ii) to raise the capital needed;

(iii) to identify the site and acquire the land;

(iv) to obtain the building and operation licences and permits;

(v) to build the plant.

Most of the developers, public or private utilities and multinational companies, being large concerns whose activities attract the public interest, give a high priority to their relations with the public and to the public opinion with respect to their policies. Hence there can be little doubt that they will make considerable effort to accommodate the environmental concerns of the local public, who, in many instances, are their customers.

The developers' top priorities still remain meeting the demand for their products and making a financial profit. The importance of regulatory procedures is that they provide the legal framework within which the developers have to rationalise and accommodate their priorities.

(i) Evaluation of the need for an MEF and evaluation of its size

Utilities have a strong interest in increasing their capacity because:

- they are usually guaranteed a profit per kWh produced;
- they obtain the capital in favourable terms with respect to other industries;
- they are not usually allowed to diversify their range of products.

The investments of public companies are controlled by the relevant authority, national, regional or local. For privately owned utilities demand for electricity and availability of capital determine the need for further capacity.

Due to the long lead time needed for building power stations there has been in many instances an overestimation of future demand which has led to overcapacity. There are also instances, during slowdowns of the economy, that national administrations encourage utilities in building power stations to help reduce unemployment, to keep an important sector of the engineering industry busy and in this way to assist the economy in its recovery and to meet the demand for electricity when this has occurred.

Multinational and other private companies acknowledge the need for further capacity on the basis of projections and evaluations of the demand for their products, on the availability of energy resources (North Sea Oil, Western United States coal, etc.), on opportunity cost, on availability of capital, and on the position of their competitors.

In determining the size of MEFs, developers follow the trend for larger installations that characterised the 1960s. Economies of scale is the justification for not considering smaller plants. This, however, becomes questionable given the advantages that smaller plants may provide, as is described in Chapter II part I.

(ii) Raising the capital

Large public companies (e.g. CEGB) must seek funds from the Administration and these can usually be obtained under favourable conditions since the risks are greatly reduced because the borrowed funds have been raised by the Central Government. Large private investments must be financed by a mixture of internally generated funds, borrowed debt capital and new stock issues. The mixture of sources used will depend on the state of the financial markets and the economic condition of the individual firms. Interest rates in any case are higher than for public companies. Normally, however, financing has not been a major problem for utilities because of the monopolistic nature of the business and the regulatory principles (e.g. guaranteed rate of return) used in most OECD countries. For other than utility enterprises the ability to raise capital will depend more on the financial condition of the company and of the money markets.

In most cases the developer does not sustain financial loss if he has raised the capital before obtaining clearance to proceed with the building of the plant. In many instances, this is not possible (e.g. in the United Kingdom CEGB, to obtain the funds from the Treasury, must first satisfy the Minister for Industry that it has obtained the necessary permits), but even if it is possible, such funds may be made available with a profit, in case of delays, as short-term loans in the money market.

(iii) Acquiring the site

When the need for further capacity is established, developers survey areas where a site may be found which meets their criteria. Their criteria are largely technical and economic:

- physical characteristics: geology, geography, topography, hydrology, climatology, availability of land...
- infrastructure: transport (roads, railways, ports, airports), communications, skills, labour, housing...
- economics: proximity to raw materials, to markets, government policy for regional development...
- legislative constraints: proximity to urban areas, emissions to the environment, water utilisation...

The developer usually acquires a site after coming into agreement with the local authorities on the suitability of the site for the proposed MEF. Although local authorities welcome, in most cases, major industrial developments there are many instances when the local population disagreed with the local authorities' decision to accept an MEF. There are other cases when the developer owned or acquired land before obtaining clearance on the part of the local government. This commitment on the part of the developer, sometimes to avoid land speculators, has led to many difficulties when the acceptance of the MEF was delayed or denied.

Following the developments in environmental legislation and regulatory procedures and the increased difficulties in public acceptability of MEFs, the developers acknowledged the economic losses associated with delays or refusal to build on acquired sites and tend to accept as an important criterion the probability that local population will agree to the siting of the MEF. To this end they have been seeking advice from consultants or carrying public opinion surveys before committing themselves to a site.

The acquisition of the site before acquiring the licences is an instance which may lead to financial loss for the developer and to acute disputes and adoption of polarised positions in disputes among interested parties. There are many cases where this has occurred either because the site was eventually found unsuitable for the particular MEF, or because the public did not accept the siting decision. As both speculation on land and delays or denials in the use of acquired land add to the overall operating costs of the developer, solutions to this problem are necessary if the siting procedures are to become more efficient. This item will be examined later in this chapter.

(iv) Permits

Apart from the building and operation licence which are the basic permits for going ahead with an MEF, the developer must acquire a large number of individual permits from different local, regional and national administrative departments. For example, a nuclear power plant in the State of Massachusetts needs 25 permits issued by seven state departments, ranging from the Certificate of Public Necessity and Convenience of the Department of Public Utilities to the Review of Buildings which may pose aviation hazards of the Department of Labour and Industry. Although many of such permits are obtained easily as a matter of formality, others need considerable time and may lead to substantial delays.

(v) Building the plant

Construction of the plant is mainly the concern of the developer but he must maintain health and safety standards of various governmental bodies and obey other rules and regulations. For most MEFs the large vessels and major equipment is purchased from private suppliers. The design of the plant and management of construction can be handled either by the developer or by a firm engaged for this purpose.

The local authorities

Local authorities have the responsibility for the use of land and local land use plans are the only ones established by law in most

OECD countries. Zoning the land and issuing permits for its use is an important source of influence for local authorities. Local tax rates on property and on company or personal income are usually determined by local authorities according to the items that a local budget has to cover. For example, in the United States the schooling system is financed by the local community while in most European OECD countries this is done by the national or regional government. As has been indicated in Chapter II part II, local tax income from an MEF is substantial for most communities and in some instances - when population is scarce - may become overwhelming. This income may be used to improve local services or to reduce the tax rates paid by the inhabitants, or both.

It has been the practice for local planning authorities to reconsider their land use plans and to rezone areas to accommodate industrial and urban development. Furthermore, utilities in many countries have the statutory authority to site their installations where they consider it will serve best the community. The local public has tended to object more strongly whenever existing land use plans were abandoned to accommodate MEFs.

Local authorities are responsible for issuing many of the permits that are necessary for the MEF. Transmission lines, use of water and electricity, sewage disposal, etc. These permits are provided by the different statutory authorities which are local or subsidiaries of regional authorities.

When the environmental concerns became evident, local authorities undertook the task to inform the public and to organise public hearings and other public participation activities and to transmit the outcome of these activities to the relevant authorities, national or otherwise. In this way they have gained experience in interpreting public feelings with respect to siting issues. They are, however, in a position of weakness because when a site is unacceptable they can hardly offer an alternative to the developer, and they have to face the often crucial dilemma of either accepting substantial environmental disruption or refusing badly needed jobs and business.

Lastly, but not the least, local authorities provide the necessary infrastructure for the construction and operation phase of an MEF as has been described in Chapter II part II.

In general, local authorities tend to be favourable to industrial development and hence to the siting of MEFs, unless traditional local industry and business are likely to be harmed, when the balance of political power within the community is likely to be changed, or when the public has strong feelings about protecting its way of life and the character of the area.

The environment protection groups

The role of environment protection groups was not widespread until the factors earlier described became important. The environmental movement may be considered an integral part of the public movement towards increased participation in the policy - decision-making process. The importance of the movement grew in the early 1960s but it was not until the late 1960s that environmental groups began to play a major role in decision-making - particularly in relation to the siting of MEFs.*

Public interest organisations encompass a broad spectrum of organised groupings that, at times, emphasize environmental concerns. All vary in their roles, functions, choices of issues, and strategies. Among the list are traditional conservation, environmental public law and other special interest organisations, all of which may organise in coalitions, informal, and ad hoc groupings. In addition to these categories of organisation, a public interest group that focuses on other than environmental concerns may also have overlapping or contiguous interests, as exemplified in alliances with labour groups and consumer activists.

Environmental public interest groups have two basic tasks in MEF siting and design issues: educating the public and intervening in public policy-making and litigation.

The public education aspect involves disseminating information, gaining public attention, organising, building and maintaining the constituency of a local group. This aspect of environmental groups involves the grassroots activism of many organisations. This is the sole function of some organisations. Diverse resources are used for organising, informing and publicising environmental issues. Writing letters to political officials, organising political demonstrations and public rallies, workshops, conferences and task forces are among the methods used.

Political and legal intervention is achieved by various strategies that require more sophisticated and specialised expertise. The environmental movement, like every other major social movement, eventually turned to the legislatures and the courts to define the roles, rules, procedures and social norms for effective participation by the public. Regulatory procedures, environmental protection legislation such as the NEPA in the United States and administrative courts in the United States and Germany, provided opportunities to the citizen and to environmental groups to intervene in siting decisions.

Environmental public interest groups are involved in two aspects of energy policy :

---

* A more detailed description of environmental and public participation groups and their activities is given in the OECD document referred to in page 5.

(i) in the formulation of some energy policies, notably the development of nuclear power, the need for more R & D and for an accelerated development of new, less polluting, energy resources and the need for increased energy conservation;

(ii) the choice of individual sites for MEFs.

For the first aspect, environmental groups have been lobbying the administrations and politicians and informing the communications media. In many instances the choice of a site was used as an opportunity for expressing in a dramatic way their policies and for attracting the attention of the public.

The large national and international environmental groups have been particularly active in opposing the development of nuclear energy. It is difficult to single out and to evaluate their role among many other factors such as the concern of the public over dangers from radioactivity, the economic slowdown, the drop in the demand for electricity, etc. which may be also considered as leading to a slowdown in the development of nuclear power. But large or small, their role cannot be denied.

With respect to siting problems, environmental groups act as catalysts in the establishment of local pressure groups. Citizens involved in public policy issues are more effective when they deal with specific issues, especially at the lower levels of government where their efforts and representation are more concentrated. Although the views of the public may be thought to be represented by government officials who are concerned with siting, experience has shown that frequently important minority opinions are not considered. Since elected officials are subject to a variety of other influences, unless the environmental and public interest views are brought specifically to their attention in a formal and systematic way they may not receive the consideration that is their due.

The public, through its organised national and local groups, is now recognised as an important participant in the siting process.

National and regional* government

National government in centrally governed countries and federal and regional government in federal states influenced siting decisions through their role in formulating and implementing energy policies, by facilitating the raising of the capital needed by the developers and by enforcing environmental legislation.

Over time, however, as concern increased over protecting other values affected by siting of MEFs and as the affected area increased,

(*) The terms "regional" government or "regional authority" are used in this Report to describe the authority - if it exists - between the national (Federal) government and the local authorities.

the involvement of both regional and federal levels of government has increased. Energy policy and security have become major issues and these are the responsibility of central governments.

Concern over environmental matters and other amenities has become increasingly a matter of national or regional concern. Changes in laws and regulations have accelerated these trends. When environmental legislation was passed in the early 1970s national (federal) and regional governments influenced siting of MEFs by setting standards and implementing and monitoring the application of these laws (clean air and clean water legislation, etc.).

The most important development in the participation of national and regional government in siting issues was the establishment by them of regulatory procedures which were to be followed when siting MEFs. These procedures fall into two distinct groups:

(i) procedures based on the Environmental Impact Statement (EIS)* and environmental protection legislation;

(ii) procedures based on land use planning and environmental legislation.

(i) Environmental Impact Statement. The first attempt to assess the environmental impacts of MEFs in a co-ordinated way was made by the Federal Government of the United States following the passage (1970) of the National Environment Policy Act (NEPA). NEPA states that:

"... all agencies of the Government shall:

... include in every recommendation or report on proposals for legislation and other major federal actions significantly affecting the quality of the human environment a detailed statement by the responsible official on:

- the environmental impact of the proposed action;
- any adverse environmental effects which cannot be avoided...
- alternatives to the proposed action;
- the relationship between local short-term users of man's environment and the maintenance and enhancement of long-term productivity;
- any irreversible and irretrievable commitments of resources...

In addition, the responsible federal official must seek help from other agencies that have special expertise, and have the EIS reviewed by appropriate federal, state and local agencies...".

Many types of MEFs are considered "major federal actions..." (notably nuclear power plants) and EISs have to be prepared for them

---

(*) Throughout the Report (EIS) is used to indicate the Environmental Impact Statement as defined in the United States National Environment Policy Act (NEPA), while the term Environmental Impact Assessment is used to indicate an assessment of impacts without defining the form it may take.

before the siting decision. However, there are MEFs for which there is no direct Federal responsibility and therefore no EIS is needed. Many of the States have passed legislation for State EISs which cover such loopholes but also produce some duplication.

Two major developments followed the passage of the NEPA: (i) the EIS became the cornerstone of the United States environmental policies, and (ii) the courts have been brought in to interpret and enforce the application of the EIS procedure.*

(ii) Land Use Planning (LUP). LUP on a wider than local basis has been adopted by countries with high population density, namely some of the European OECD countries, particularly during the years of industrial reconstruction and urban development that followed the Second World War.

LUP procedures are carried out, in most instances, within administrative departments with limited participation of the public or other groups. Compliance with air and water standards when these laws were passed in the late 1960s and early 1970s was left to central (federal) administrative departments, hence the large number of permits that utilities have to obtain when applying for a building licence. Emission and ambient concentration standards have affected siting decisions only in highly industrialised areas where there were possibilities that they could be exceeded.

Modifications in LUP laws in some Member countries have emphasized the role of central and of regional government in the siting of MEFs. For example, in both the United Kingdom and the Federal Republic of Germany recent changes in LUP legislation placed greater responsibility with the regional governments, while in the United States the Coastal Zone Management Act (1972) has, for the first time, provided some incentives to the State's government for LUP.

C. CURRENT DEVELOPMENTS WITH THE SITING PROCESS

During the mid-1970s many OECD governments realised that the limited number of participants, the current regulatory practices and the social, technical and economic development earlier described were making the siting of MEFs increasingly more difficult.

They also foresaw that given the current social trends the situation might worsen in the future. To solve their problems most governments turned to their regulatory procedures, thus indicating that if the larger number of participants and the multitude of their criteria had to be accommodated, only an improved regulatory procedure could provide the new rules needed for decision-making. Thus from 1972 to this day many governments have proposed, passed, or

* The EIS procedure and the experience acquired in the six years of its application is discussed in the OECD Document referred to in page 5.

begun to implement legislation aiming at improved procedures for siting major industrial installations. This new legislation - which in many instances provides for broader participation, dialogue and consultation among interested and affected groups as well as for considerable reallocation of responsibilities in the process - is in its early stages of application and appeared while this report was being prepared. The importance of regulatory procedures has been singled out by the Secretariat at the beginning of this project as one of the key factors for siting MEFs and considerable effort has been devoted since into studying ways by which regulatory procedures could respond better to the changing needs in Member countries.

Before presenting the Secretariat's findings and proposals, a review of a number of recent developments in regulatory procedures may help to illustrate the extent to which Member countries have been active in this area, the trends that seem to emerge in their thinking and the background on which this report's proposals are based.

Examples at the national level

In the United Kingdom, The 1971 Town and Country Planning Act establishes the structure plan which represents a major departure from the previous system which is intended to look about 15 years ahead. The structure plan is a written statement illustrated diagrammatically which states and justifies the choice of policies. Structure plans are prepared by County Planning Authorities in England and Wales. In Scotland, Regional Authorities carry out this task, but only where such plans are deemed necessary. Local Plans are normally prepared by district planning authorities in England and Wales and in Scotland. The preparation of the structure plan gives the opportunity for an early consultation and evaluation of potential sites by the regional authorities bodies, the local authorities and the CEGB.

The use of this new Act is illustrated, for example, in the Structure Plan for South Hampshire (approved, after verification, in 1977):

> "Arising from both regional and local studies of the most suitable means of meeting the increasing demands for electrical power, a need has been established for certainly one, and possibly two, new power stations in the Plan Area before 1991. Joint studies by the CEGB and the Local Planning Authorities have established that of six alternative sites studied in detail, a site at Fawley would be suitable for the first additional station, probably required in the early 1980s. Uncertainties in technical developments and future fuel policy make it impossible at present to establish the most suitable site for the second station. It has not been possible to rule out any of six sites as physically unsuitable for power station construction."

And later in the General Proposals:

... consultation and joint study has already led to a number of proposals. Of significance to the Structure Plan are the decisions of the Local Planning Authorities to:

(i) accept the need for the construction of a second power station... for completion possibly in the early 1980s, and to ensure that no development incompatible with such use is permitted;

(ii) bring to the attention of the CEGB any proposals for development on or in the vicinity of the five other possible power station sites... and to take their views into account before making a decision.

The Federal Government of the Federal Republic of Germany has announced*, in its 1973 Energy Programme, measures designed to ensure the timely identification of suitable sites for the development of an adequate energy supply. This is repeated in the 1974 Energy Programme where it is defined that a 1985-1990 horizon for sites for MEFs should be established. This again is stated in the regional development report of that same year. Furthermore, evaluation criteria for nuclear power plant sites were established.

In the 1976 Environment Report the following statements are made:

"... suitable sites for supraregional industrial facilities, particularly for energy-producing facilities, must be identified. The Federal Government in co-operation with the Länder shall work out a scheme for planning and securing sites... for supply and disposal facilities in the energy field within long-term plans for regional development. The participation of the affected population should be ensured in such schemes.

This work programme comprises the development of general and assessment data for an early examination of sites, a survey of siting plans of the Länder, the reaching of agreements between Länder for borderline projects or with neighbouring countries and the long-term EEC siting policy. The Federal Government will co-operate in these tasks with the Länder responsible for siting and licencing to ensure a uniform Federal siting policy".

United States: The Coastal Zone Management Act

Recognition that land use decisions are reflected in a wide variety of other policies affecting the national welfare has resulted in numerous attempts to enact limited land use legislation at the Federal level in the United States.

(*) For a more detailed description of developments in the Federal Republic of Germany, both at the national and regional level, see the OECD document referred to in page 5.

However, the only major Land Use bill enacted has been the Coastal Zone Management Act of 1972 (CZMA). The Congress concluded that the coastal zone needs urgent attention because large metropolitan areas with their suburban sprawl had blotted out great stretches of shoreline and heavy industrial complexes had entered the coastal zones and utilised their natural resources. It was further recognised that the CZMA had to be managed in a way which would be compatible with any broader land use legislation which might be enacted in the future.

The CZMA provides States' administrations with incentives to develop and implement coastal zone land use plans in the form of grants available to them under prescribed criteria for developing management programmes, for administering these programmes and for acquiring estuarine sanctuaries to be used as field laboratories in which to gather data concerning estuaries.

The stated policies of the CZMA are as follows:

(i) to preserve, protect... the resources of the nation's coastal zone...

(ii) to encourage and assist the states to exercise effectively their responsibilities in the coastal zone through the development and implementation of management programmes...

(iii) for all Federal agencies... to co-operate and participate with state and local governments and regional agencies in managing the coastal zone;

(iv) to encourage the participation of the public... in the development of CZM programmes.

The Senate began an attempt in 1975 - without success yet - to enact a Land Use Bill very similar to the CZMA. The Bill would have a major impact on MEF siting since it proposes among other things that "the siting of energy facilities be integrated with state land and water resource planning and management, the state land and water resource programme, and any coastal zone management programme..." and that "provisions be made for states to establish, operate and fund energy facilities planning programmes". It further proposes an Energy Facility Planning Process which shall, among other things, "... identify intermediate and long-term energy demand, resources, conservation programmes, ... and facilities, ... recommend energy conservation measures, identify energy facilities necessary to meet projected energy needs, evaluate the economic, social and environmental consequences of developing and operating projected energy facilities...".

In France, considerable developments took place recently in the siting of nuclear power plants. Previously, the siting of power stations was being carried out by Electricité de France (EDF), the Ministry of Industry and the local authorities. When the oil price crisis occurred, and the Government realised the nationwide anxiety the installation of 150 nuclear power stations by the year 2000 was creating, the administration adopted DATAR's (the Government LUP authority) suggestion for the participation of regional authorities.

Thus, the consultation (concertation) of the Regional Councils* took place in 1975. The Regional Councils had to express their opinion as to the acceptance or rejection of one or more sites which were chosen by EDF and the Ministry of Industry, and/or to indicate alternative sites.

Some of the Councils took a decision after calling a referendum of the regional population, but most did so by consulting local authorities and municipalities. The result of the "consultation", put in a nutshell, was that some regions agreed, out of hand to the siting, some reserved their opinion till the studies on the environmental and other impacts which are identified as necessary for the decision-making process were completed and evaluated, and some expressed outright opposition. Thus, EDF and the Government had a good indication of how to proceed with their nuclear programme.

Despite the evident drawbacks - nuclear authority of the regional councils, lack of organised public participation, absence of environmental impact assessment - the "consultation" has been an important development in the siting procedures of France which emphasized - as never before - the political nature of the process, the importance of regional and local authorities and the role of LUP Administrations.**

In the Netherlands the planning of projects for electricity supply took place almost entirely within the electricity sector. There had been no scope for assessing the various interests involved owing to the absence of a long-term policy for the electricity sector. A major recent development was the adoption of a long-term policy and of a planning instrument (structure plan) for reserving space for major works of infrastructure, for detecting conflicts with the policy sectors at an early stage, and for enabling timely, clear-cut and well founded decisions to be taken.

Electricity supply is an important part of that plan. The electricity supply structure plan contains a list of sites which have been accepted by the Government as possible sites for power stations of more than 1000 MWe capacity. In most cases these sites have not been determined accurately. If at a later date, after the objectives in the structure plan have been assessed, the need for building a power station should arise, the list contained in the

---

(*) Regional Councils are composed of the senators and representatives of the Region, the delegates from each local authority of the Region and the delegates of the cities of the Region.

(**) For a detailed description of France's policies in siting MEFs see the OECD document referred to in page 5.

structure plan offers a framework within which a choice is possible.

The structure plan becomes the subject of interdepartmental consultation and of consultation with the provinces. It is made public and every citizen and group of citizens is given the opportunity to make his views known before the plan is finally adopted by the government. The possibility of giving it a legal basis is being considered, for example by mentioning it in the Physical Planning Act and in the Electricity Act.

Projects featured in the structure plan should first be assessed in a regional plan as to their acceptability, and specified in detail. The regional plan is the appropriate framework for this, because all physical developments are considered in it and all interests weighed against one another. This assessment is done following a procedure of participation and petitioning by a democratically elected body: the provincial executive. This procedure may establish that implementation of a project is unacceptable according to the priorities set. Then the project could either be scrapped, if there is agreement by the national planning authority, or be reconsidered if the national interest is thought not to have been taken sufficiently into account by the regional authority.

Examples at the regional level

The Land of Baden-Württemberg. Within the framework of the Federal Land Use Law the Land of Baden-Württemberg produced a "specialised development plan for siting MEFs", which sets aside 14 sites for MEFs and prohibits any other uses of land which may prejudice against building power stations there. This plan was prepared by the Land Planning Board, the Federal government, the neighbouring Länder and the neighbouring countries (France and Switzerland).

The Land of North Rhine Westphalia. In the light of the Federal Law on Regional Planning, the state of North Rhine Westphalia, when preparing its 6th Land Development Plan, has identified sites for MEFs including nuclear power plants and has examined the conformity of those sites with the Federal Laws for the protection of the environment (Emission Control) and for protection from radioactivity. When this is completed other regional administration departments will examine the proposed sites from their point of view. As a result of the examination, the selected sites are included in the Draft Plan which is being sent (April 1977) to the district planning authorities and to the appropriate communities for their opinions. After assessing these opinions the Land (State) Planning Authority will draw the Final Plan and submit it to the Cabinet for approval.

Once it is approved and published, the district planning councils are required by the Land Government to incorporate the Plan objectives into district development plans. The local authorities should then draw the necessary local land use plans, a necessary prerogative for the claim to the proposed sites. As the local authorities have participated in the drawing of the plan it is expected that only in a few cases the Land Government will have to issue planning orders to the communities to obtain the necessary sites. In this way it is expected that sites for MEFs will be identified and made available to the developers in a much simpler way.

The State of California established in 1975 the California Energy Resources Conservation and Development Commission which has been given a mandate which includes: ...determining the need for new power plants, evaluating and certifying proposed designs and sites...

The State of Maryland has enacted legislation which establishes an Environmental Trust Fund for financing a power plant siting and research programme. The funds derive from an environmental surcharge per kWh of electric energy generated in the state to be paid by the consumers.

(a) The Public Service Commission of the State assembles and evaluates annually, the long-range plans of Maryland's public electric companies regarding generating needs and means for meeting those needs. It puts forward annually a ten-year plan of possible and proposed sites, including associated transmission routes for the construction of new electric power plants and extensions of existing plants.

(b) A preliminary environmental statement on each possible and proposed site, including associated transmission routes, is prepared.

(c) The State acquires a sufficient number of sites to satisfy the expected requirements. Site selection is based on existing research findings that show the site is desirable for power plant construction. The State holds the property and may not permit its temporary use for any purpose which might logically be expected to impede its prompt availability when needed. Seventy-five per cent of all revenues the State obtains through temporary use of sites shall be deposited in the fund. The remaining 25 per cent shall be paid to the county in which the site is situated. The electric utility may purchase the site, or lease it on a 99-year lease. The purchase price shall be the fair market value of the site.

D. BASIC PRINCIPLES FOR REGULATORY PROCEDURES DEALING WITH SITING MEFs

This and the following section have been written with the knowledge that for some Member countries constitutional and administrative constraints may make impossible the adoption of some of the proposed measures. These procedures and measures therefore have to be interpreted as guidelines for possible future action on the part of Member Governments given their constitutional and administrative constraints.

After considering:

(i) the historical developments that led to the complications in the siting of MEFs in OECD countries and their needs in MEFs as projected for the next 15 years (Chapter I);

(ii) the flexibility that pollution control technologies and, in particular, the adoption of new technologies and of radical measures may provide in siting (Chapter II, part I);

(iii) the economic and social benefits that may accrue for the local population and the danger that lack of planning may, on the contrary, adversely affect the society and economy of a community (Chapter II, part II);

(iv) the demand by many groups to participate in the process, the elaborate and in many instances contradictory criteria adopted by these groups; and

(v) the emerging trends in the thinking of Member governments.

Four principles emerge as possible basis for a framework for regulatory procedures for the siting of MEFs:

1) the planning for MEFs should be based on long-term policies integrated to national and international energy policies;
2) the siting of MEFs and the assessment of their environmental impacts should be carried out within long-term land use plans;
3) regional authorities should undertake the major responsibility in the siting procedures;
4) public participation should be encouraged and incorporated in all stages of the siting process.

(a) Integration of energy policies and siting policies

The planning for MEFs should be based on long-term policies integrated to national and international energy policies

Following the oil price crisis, the availability and security of supply of energy resources became a matter of the utmost importance for most OECD countries. Central (Federal) governments have to decide on the energy resource formula that is most suitable at any particular time horizon, given the economic and political constraints. Thus the developers' planning departments cannot decide by themselves on the types of MEFs needed on a long-term basis as they usually did before 1973.

Given the large size of MEFs, the long lead times required for such plants to be built and be put on stream and the increasing opposition of the public to accommodate them, it had become apparent that the availability of sites will be a real constraint to the implementation of energy policies, and as such, it has to be taken into account earlier in the planning process.

The adoption by Member governments for long lead time procedures for the siting of MEFs appears to be an important factor, both for the reduction of adverse environmental impacts and for the smooth implementation of energy policies. Such long lead times not only allow for comprehensive evaluations of environmental, social and economic impacts but they may also allow for the local public to be prepared for accommodating these impacts. If the objections of the public, as indicated earlier, are objections to change, then the elimination of the element of surprise will reduce the strength of the reactions by spreading them over a longer interval of time and by allowing both the authorities and the public to better understand the issues and to decide accordingly.

Energy policy is closely associated to economic policy, foreign affairs and trade. It is, therefore, the responsibility of the national (federal) administrations. The responsibility for siting MEFs is, on the contrary, distributed among national governments, regional and local authorities and developers who could be public, private or multinational companies. The process of siting is also subjected to considerable pressure by environment protection groups and the public. Thus the siting of MEFs cannot be decided at any particular level of government.

For national (federal) governments to ensure that the siting of MEFs will not be a constraint to the development of their long-term energy policies calls for a co-ordination far beyond the consultation level, of energy and economic policy departments on the one hand, of environment and land use planning departments and of the regional authorities on the other. There can be little doubt as to the fact that the time horizon for energy policy and for land use planning must be similar.

Apart from the problems associated with siting energy facilities of the types now in use, the coming decade promises some radical changes in the sources of energy. Energy R & D has been given much attention in the last few years, and as most energy R & D projects. are very large, call for the commitment of considerable resources for long periods of time, and if successful, may need facilities with considerable demands for land, water and other natural resources or bring to bear on the environment particular impacts, it is necessary that the siting constraints for such facilities are evaluated when decisions are taken to invest into energy R & D projects. Otherwise, the sheer size of committed R & D resources during development may be used as argument for the acceptance by the public of controversial or polluting facilities and in this way help to prolong the attitude of confrontation between interested groups.

(b) Environmental impact assessment within land use planning

The siting of MEFs and the assessment of their environmental impacts should be carried out within long-term land use plans

LUP and EIS procedures, as practiced for the siting of MEFs, seem to be complementary in the sense that what one lacks the other provides. Thus LUP is carried out in most countries with little assessment of environmental impacts of either the plan as a whole or of individual developments within the plan, while EIS procedures initiated by an application for a construction licence lack the long-term dimension which characterises LUP.

LUP legislation, with few exceptions, does not deal with the protection of the environment, apart from protecting historical sites and buildings, amenity and recreation areas. It deals, at the national and regional levels, mainly with the zoning of land, so that a balance is kept among users of land (agriculture, industry, urban centres, etc.) to serve the economy, and at the local level, to assist land development, to maintain and improve the price of land, to accommodate the local population and to provide transport, communications and other services.

EISs on the other hand have shown during the six years of their application that, with respect to the siting of MEFs, they have led industry and administration to co-operate in the evaluation of adverse environmental impacts on possible sites, encouraged industry to apply environmental and public acceptability criteria when choosing a site, opened the way to the public to participate in the process, and enlarged the perspective of the decision-making process within the administration. Agencies, however, have tended to produce EISs that are too long, too descriptive and not sufficiently analytical, being prepared on an ad hoc basis, they bear little relation to long-term social, economic, energy and land use plans, they have had little success in evaluating alternative sites unless the developer owned more than one, and they have been used to evaluate only a few of the broad agency policies and the alternatives to these policies.

Important as the environmental impact assessment of an MEF is, advantage of its beneficial effects cannot be fully gained unless it is carried out within a long-term LUP. Some of the drawbacks of EIS, identified earlier, may be eliminated in this way. Examples:

(a) the long descriptions of geography, topography, flora and fauna for each and every major installation sited in the same area may be avoided if this is taken care of within a regional LUP;

(b) alternative sites for MEFs may be considered - and found - within a plan which has established where land can be used for industrial development, without this leading to unacceptable distortions of land values;

(c) if regional authorities had a major say in the development of LUP, the environmental assessment can be made in the light of long-term social, economic, energy and land use projections and not only with respect to current situations and the immediate future.

Many OECD countries have examined the EIS procedure. Australia and some Canadian provinces have adopted it as such and many others have taken measures, or are considering taking them, to introduce an assessment of environmental impacts prior to going ahead with particular types of development. Thus the United Kingdom examined the way an environmental assessment may be introduced in the existing siting procedure. France's law on the Protection of the Environment (1976) calls for an assessment of environmental impacts, land use effects and other nuisances by the developer prior to an authorisation of a building licence for any major development. The Federal Government in Germany decided on the principles applicable for reviewing the environmental impact of measures taken by the Public Authorities of the Federal Republic. Lastly, the EEC Commission is examining a proposal that environmental impacts of major industrial installations are evaluated before a building licence is issued.

The adoption of an LUP system on a regional or national basis may meet with considerable difficulties in some OECD countries because of constitutional, philosophical or administrative matters, e.g. when the citizen's rights on property are affected, when local authorities alone are vested with the right to issue land use plans, when the budget for developing such a system is considered very high, etc. It does, however, seem that if the siting of MEFs is oonsidered as an integral part of long-term energy policies, only land use planning on a comparative time horizon can lead to the coupling of siting to energy policies.

(c) Distribution of responsibilities among levels of government

Regional authorities should undertake the major responsibility in siting procedures

If national (federal) government is best placed to undertake the formulation of long-term energy and siting policies, regional government seems to be best placed to identify, evaluate and secure sites for MEFs on a long-term basis through regional land use plans.

It may seem that Federal States will have less difficulty in adopting such a system because regional authorities are vested with the responsibility over land use. This is not, however, always the case because some of the responsibilities which may be reallocated to regional government are those of local authorities for both centrally governed and federal states.

Regional authorities seem, in theory at least, to be better placed to shoulder a major part of the responsibility for LUP and for siting MEFs. Regional authorities, unlike local authorities, cannot avoid siting some MEFs within their geographical boundaries. They can, however, offer alternative sites to developers which will suit local preferences and needs for economic development of the region as a whole. In many instances energy resources have to be managed on a regional basis and any national siting policy has to take into account the economic and employment opportunities that energy development offers to the region. With or without energy resources, regional authorities have a better understanding than local governments of national energy needs and policies and are likely to consider local preferences and local interests and national and regional energy needs in a more balanced way.

But if regional government is better placed with respect to taking decisions on siting MEFs, it is necessary to qualify the statements of the previous paragraph in terms of size and administration of the "region".

The differences in size, population and population density among a Canadian province, a Swiss canton, a French department, etc., are large as indeed they are among states, e.g. in the United States - New Hampshire, Montana, New York State, Texas, etc. Hence, the siting of an MEF is bound to have varying significance to the regional authorities as it will affect varying segments of the population. Furthermore, a small, less industrialised region would expect few MEFs, hence the siting will be a minor activity for the regional authorities, while a large, or a heavily populated and industrialised region may require substantial financial and administrative effort on a permanent bases.

Attempts have been made in some OECD countries to devise another level of authority, sometimes within a region, sometimes bringing together a number of regions, to deal with land use plans. An example is river valley authorities which found themselves involved in siting issues by implication. However, river basin authorities lack the stature, planning resources and legal power to undertake a major role in siting MEFs as their role is limited to the determination of the overall allocation of water resources in the basin.

Given the realities of local and regional politics, the political power that is associated with land use and development planning, the constitutions of OECD countries and the fact that siting MEFs, however important, cannot be permanently a major activity of an administration (be it local or regional), the creation of an administrative structure for the siting of MEFs at any other level than the existing levels of government is not advisable. The disadvantages, therefore, due to differences in size and population of OECD countries' "regions" may be accommodated with less difficulty than if the creation of some new type of more homogeneous "region" was attempted. Much more so if we consider that discrepancies among existing "regions" are not only due to size and population but also to the presence or absence of energy resources, the extent of industrialisation, transport facilities, geographic location, etc.

As the problem associated with siting MEFs may be reduced by administrative procedures, as the public demands a larger share in the decision process, as the public affected by an MEF tends to live in more than one community, as the environmental impacts are spread over wider areas, it does seem that regional government should play a major role in the siting of MEFs in co-operation with local authorities and national (federal) government.

But to play such a role, be it through land use plans, through the creation of special departments for this purpose, through the co-ordination of local authorities and other bodies which have statutory authority over aspects of siting, etc., regional authorities should be given, or allowed to acquire, the necessary institutional, technical, human, financial and administrative means which will ensure that the task will be carried out in an efficient way and, in particular, that a wide spectrum of criteria will be established and used for siting decisions.

(d) Public participation

The greater the extent and impact of a proposed development the greater the effect it is likely to have on the local population, particularly for those living in close proximity to the site. In the past, in most OECD countries, persons or bodies other than the developer and the local authority have had no right to be heard before the decision was reached; their views were assumed to be expressed adequately by their local elected representative. Obviously, the elected representatives can only vote one way and have no accurate means of assessing the wishes of their constituents and, in any case, are under no legal obligation to reflect those views. In recent years it has been realised that those affected, either directly or indirectly, should have a positive method of participating and there is now a legal requirement in most countries that all so-called "bad neighbour" developments are advertised in the local press and that a decision is not reached until all objections or comments have been taken into account. Although there are many cases where local opposition has been instrumental in stopping or altering significantly proposed projects, only nuclear power plants among MEFs have been consistently subjected up to now to such opposition.

It has been the experience in some OECD countries, notably France, Germany, the United States and United Kingdom, that a nationwide reaction to nuclear power followed the announcement of major commitments to nuclear power and the need for a large number of sites. It seems that the announcement acted as a catalyst, bringing together all aspects of opposition and confusing the issues. Had the policies (including the number of sites) been developed with broader participation of the public and interest groups, it is possible that the programmes would not have met with such stiff opposition - either because the policies might have been changed (as they eventually were) or because the public might have better understood the reasons for the policies.

Furthermore, some local communities have accepted the siting of nuclear power plants with little or no opposition, some convinced that this was just another one of the many risks with which we live every day, others because of the economic and social benefits which accompany the siting, and others because of indifference. This implies that previous experience with siting in a country and public attitude surveys around potential sites for nuclear power stations may be useful to identify the particular concerns of citizens and to determine the receptiveness of the public. Such evaluations may be considered before the developer is encouraged to proceed with his application.

Public participation is not a panacea to wise selection of sites. It can, however, help in making the choice by reflecting the public's preference; i.e. by reducing the social adverse impacts. On the other hand the public (local, regional) may take a negative attitude to any development, if it is not given the opportunity to express its opinion and have it considered by the authorities.

Although many ways of public participation have been tried so far, in most of them the local public is brought in when the regulatory procedure is nearly finished. This is justified by assuming that regulatory authorities will remove most objectionable impacts and will present to the public something legal and as "clean" as possible. But the local public objects sometimes for purely "local" matters, e.g. two of the early controversies over the siting of nuclear power plants in the United States had nothing to do with concern over safety but with the choice of sites which were close to local "beauty spots".

The opposition of the public is not confined only to local problems. The experience, described in many parts of this report, is that the public objects also on a national, even an international basis. Nuclear energy is again the most evident example, but by no means the only one. Threats to national parks, to particularly beautiful natural spots or historical sites have been fought on a nationwide basis. The opposition to a shipbuilding yard in the South West of Greece which was welcome to the local population because of the employment and business opportunities it offered, was vehemently fought by environmental groups and the public on the grounds that a historical site (the Palace of King Nestor) and the Bay of Pylos would be adversely affected.

It does seem, therefore, that public participation should be integrated in the decision-making process for siting MEFs at all levels of government and not only at the local level, so that the public can express its concerns at the appropriate level of decision.

A solution to this problem which has been tried in the past in some instances with respect to nuclear power plants, was to put the question to a democratic vote on a regional basis. The experience with these referenda has shown that the public has an open mind regarding nuclear energy. In the referenda held in six states in the United States, the public voted overwhelmingly in support of nuclear energy (58-70 per cent). In the two referenda held in France, one region voted for and one against the siting of nuclear power plants.

E. A FRAMEWORK FOR A PROCEDURE FOR SITING MEFs

The application of the principles described in the previous section may be ensured by the adoption of a framework for a procedure for siting MEFs. Such a framework should ensure that policy decisions taken at the national level are pursued in a compatible way by regional and local authorities as increasingly site specific decisions are being reached.

The fundamental differences which exist among OECD countries with respect to property rights of the citizen, land use planning systems, environmental impact assessment procedures, federal or centralised government, the right of the citizen to challenge the administration in the courts, etc. seem to pose unsurmountable constraints to the adoption of any common procedure in siting MEFs. Although these and other differences may hamper the adoption of a common detailed procedure, there are also many similarities among OECD countries which tend to indicate that a common framework procedure may be acceptable by many of them. The common characteristics of most energy technologies, the similarities in the structure of administrations, and the many common characteristics found in the new procedures proposed or in force in OECD countries are among these similarities.

As mentioned earlier, although energy policy, being closely associated with foreign and economic policy is, primarily, the responsibility of national (federal) governments, the responsibility for siting MEFs is distributed among national, regional and local government.

The process of siting MEFs may be designed as a process in stages closely related to each other, the location of the site becoming more specific as the responsibility for the decision passes from the national, through the regional, to the local government.

Thus, the siting of MEFs may be considered as a three-stage process where:

(a) The first stage (responsibility being mainly of the national or federal government) seeks consensus agreements over long-term energy policies so that they are not opposed by interested parties or the public, and may ensure that the siting of MEFs will not act as a constraint to those energy policies.

(b) The second stage* (responsibility being mainly with regional authorities) seeks to identify and secure, within long-term land use plans, possible sites for those types of MEFs which will be necessary for the implementation of long-term energy policies, to evaluate the potential of the physical and social environment to absorb the expected impacts and to seek the acceptance by the public of such MEFs.

(c) The third stage* (responsibility shared between regional, local authorities and the developer) seeks to ensure that once a particular MEF is needed the site which will be selected is in accordance with long-term land use plans, that the social, economic and environmental impacts of the technology used at that particular moment in time are evaluated, that the developer and the local authority provide the necessary infrastructure to reduce adverse social impacts, and that the developer is assisted with the licensing procedures so that the project is not delayed.

The first stage

At that stage, possible environmental impacts expected from the implementation of different options of energy policy may be evaluated as well as the capacity of the environment, on a national and regional basis, to absorb those impacts. New energy technologies for which R & D resources are to be committed may be also evaluated from the environmental and public acceptance point of view.

To achieve a smooth implementation of national energy policies, energy departments should co-operate with environment protection and land use planning departments, with regional authorities and developers when formulating energy policy.

As indicated earlier, this stage should deal with:

(i) types of energy resources and extent of their development;

(ii) siting issues.

With respect to types of energy resources and extent of their development, technical, financial, economic and environmental issues have to be dealt with. The nuclear energy issue is perhaps the most important. Should a country go nuclear? What percentage of electricity production should be nuclear? Will other nuclear cycle facilities

* With respect to nuclear installations the participation of the central government in all three stages is necessary and ensured by law in most OECD countries.

be necessary? What will be done with the nuclear waste? But nuclear is by no means the only problem. An increased use of coal poses important problems of transport, of air emissions and their control, etc. Energy R & D programmes and priorities are again a thoroughly disputed issue. For example, the allocation of large R & D resources in nuclear research has been criticised as responsible for the limited progress in the development of alternative energy resources.

It has been mentioned earlier that sometimes national environmental groups object to siting decisions not because the site itself is objectionable, but because they are concerned for the development of particular energy policies (e.g. nuclear energy, the breeder reactor, etc.). The point was also made that the demand for participation reflects the desires of the public that are not effectively represented or expressed in the political forum.

The democratic process does not allow for the participation of one group in the absence of others. Hence the participation of national environmental groups which, although acting outside the "establishments" have aroused the interest and support of politicians and political parties, may be balanced by the participation of labour unions and by the representatives of industrial and business interests which may be responsible for eventually implementing, or may be affected by, specific energy policies.

It is difficult to consider such activities at a level other than that of consultation. And any participation and agreement in policies cannot guarantee the acceptance of these policies by the population at large. But, as the siting of MEFs has become a political process, a wide participation of groups involved at the earliest possible moment could lead to rationalisation and facilitate compromise.

With respect to siting issues, it is important that at this stage energy, economic, land use planning departments, the developers and regional authorities are satisfied that sites for the MEFs which are needed to implement energy policies will be available. To this end, land use planning, regional authorities, and developers should decide to make preparatory plans for MEF sites given the regional development plans, employment patterns, physical environment, financial, technical and economic constraints, as well as the capacity of the environment to absorb the adverse impacts and the acceptance by the public of these installations. Energy R & D projects have also to be examined with respect to the siting problems they will pose, if successfully completed.

The second stage

To give substance to governmental energy policy it is necessary that MEF sites are identified and secured on a long-term basis.

Regional authorities seem to be best placed to undertake this task. This has to be done within a regional long-term land use planning system. Regional land use plans may be produced by integrating local plans, by issuing directives or by providing a general framework for local authorities to tailor their plans accordingly. It seems, therefore, necessary that regional authorities' Land Use Planning Departments are given the necessary financial, institutional, human and technical means to undertake the task of identifying and securing potential sites for different kinds of MEFs.

In doing so they may assess the probable environmental, social and economic impacts that may result if an MEF is to be sited, the capacity of the environment to accommodate those impacts and the public opinion of neighbouring communities and of the population of the region as a whole.

This would be a planning procedure.

Regional authorities are singled out as the most appropriate to carry these responsibilities for the following reasons:

With the increase in the size of MEFs, their economic significance as well as the social and environmental impacts are felt far beyond the limits of the local community.

A local authority, depending on the attitude of the local population, may create environmental "black spots" by granting licences for industrial installations, or it may reject any application to safeguard traditional local industry or the price of land. It has very limited flexibility in offering alternative sites to the developer, and hence it may find itself under a "leave or take" deal.

On the other hand, central (or federal) governments which carry the main responsibility for the first stage of the process may find it difficult to identify probable sites and evaluate local or regional opinion. Thus, local opposition to a site may be misinterpreted as opposition to the government's economic and energy policies.

The identification of potential sites for different kinds of MEFs requires the study and determination of a large number of characteristics that are of importance to both developer, local authority and the public. Geology, hydrology and many of the characteristics described earlier as criteria for the choice of the site have to be determined at that stage. A substantial part of what is included in an EIS or what is needed to prove that environmental legislation is not going to be abused has to be determined at that early stage. It is also important that this is done in co-operation with developers or their associations and the local authorities. The public should be informed about the possible use of that site, be invited to express its opinion, which in turn should be given the utmost attention. In this way, the element of surprise in the change of the environment may be eliminated.

Once potential sites have been identified, the question remains of how they can be secured till they are needed, if needed at all. Experience in OECD countries can be found in the Federal Republic of Germany, where the Constitution allows that land qualified for public utility cannot be used by the owner in any way which may jeopardise the use for which it was designated. Another example is the State of Maryland in the United States which has established a special fund used to buy potential sites, clear them through the regulatory procedures for nuclear or conventional power stations, and make them available to utilities whenever they consider building a new power station. In the meantime, any income derived from the use of this land is put in a special fund to be used for buying additional sites.

Securing the sites without undue loss to the public is one of the most important parts of a siting policy, that has to be studied within countries in relation to their Constitution, legislation and property laws.

The third stage

This stage is the most important in terms of commitment of resources. Local planning authorities, the developer and the local public are the important participants at this stage. There is every reason that the stage is kept as short as possible. Despite the two first stages and the agreements that may have been reached among interested parties, the application of a developer to build an MEF may produce the same reactions on the part of the public as if nothing had happened before. This is, however, only a possibility, so that regional and local authorities should ensure that once the application has been received the least possible deviation from previous agreements is allowed.

Regional and local authorities may ensure that the sites proposed by the developer lie within the areas identified and secured as possible sites. In this way, most of the possible environmental and social impacts have been already identified and assessed, while the public has been prepared to expect such a development.

An evaluation of some environmental impacts, associated with the particular technology at that moment in time, may be necessary to complement the more general assessments made at the previous stage. This evaluation should reflect closely the apprehensions of the local public and should be specific to a restricted number of key issues. The issues may be agreed upon by co-operation among the developer, the local authority and the public.

At this stage, the developer and the local authority may evaluate the social and economic benefits and disbenefits and, in the light of the public's attitude and the expected environmental impacts, strive to reach a compromise in accommodating the MEF. An important item in this aspect is to agree on the financing of infrastructure to accommodate the impact of the construction phase of the project.

To facilitate the task of the applicant it may be found useful that regional and/or local authorities act as co-ordinators in obtaining for the developer the multitude of permits that are necessary to proceed with building an MEF.

F. THE SHORT-TERM INTERIM SITING PROBLEM

If Member countries wish to adopt those parts of the procedures proposed in the previous sections that are useful for their particular set of siting conditions, it will require enactment of new legislation, at least in many countries. For most countries making changes in legislation is very time consuming and the outcome of the debate on proposed new laws is often uncertain.

During this interim period there will be situations, however, in which new MEFs must be sited if national energy policies and goals are to be achieved and if adverse effects on other national programmes are to be averted. For most countries the number of such urgently needed facilities is currently relatively small (Chapter I). This is the result of the world-wide decline in economic activity and of the much higher prices for energy than were prevalent before 1974. Both of these factors have reduced expected energy demand to much lower levels than had previously been projected for the future. For example, for the Federal Republic of Germany projections of the nuclear capacity required in 1985 have recently been reduced from 45,000 MW to 24,000 MW. The new projections of reduced energy demand in all OECD countries have been reflected in a reduction of the number of MEFs that will be needed in future years. This has caused some relaxation in the urgency of the MEF siting problem, especially in the immediate future.

Because of the reduced number of plants that must be sited quickly it should be possible to accommodate their siting on a case-by-case basis during the interim period when any necessary new legislation is being considered to provide a more general, permanent and improved siting procedure.

For these relatively few case-by-case situations, interim procedures could consist of (a) establishing the need for a plant of a specified capacity for a particular year consistent with national energy policy, (b) identification of rate controlling step in the siting process, (c) estimate the amount of time that must be saved over that of using normal siting procedures if the facility is to be operating in the year that it is needed, (d) identify means that may be used to reduce the siting time that require changes in administrative procedures or regulations that may be modified by the regulatory body (without new legislation), (e) select the measure or measures that will permit the siting timetable to be met and that would result in the minimum adverse effects, and (f) provide an opportunity for a public hearing on the merits of modifying the standard regulations.

In the event that useful measures identified would require changes in legislation, adequate relief might be provided through variances that might be sought from the institution legally able to grant them. Opportunity should be provided for public participation in the action in which a variance is sought. The criterion for granting relief would be the balance between the adverse effects of not being able to provide the energy supplied by the MEF and the nature, importance and magnitudes of the adverse effects that granting a variance would cause.

G. CONCLUSIONS

The siting of MEFs is an important issue in many OECD countries, as manifested by the substantial effort put by Member governments during the last three to four years to provide solutions for the many problems posed by them. The political overtones of the siting issue - which used to be primarily a technical and economic one - appeared during the last few years, as a result of high rates of industrialisation, increase of the size of individual installations, the environmental protection movement, fear of nuclear power and the unbalanced distribution of environmental, social and economic costs and benefits among groups of citizens.

Energy policies and siting policies

The siting of MEFs is an integral part of national energy policies and as such it is a long lead process. As technical, economic and political factors may become constraints to the implementation of energy policies, environmental and sociological factors may also become constraints through their role in the siting of MEFs.

(i) It is, therefore, necessary that administrative bodies responsible for the development of national energy policies include representatives of the departments of environmental and of social affairs as well as representatives of regional authorities.

(ii) It is also necessary that such policies are elaborated in close co-operation with representatives of national environmental groups, of labour unions, of industrial federations and of the other groups that can affect or be seriously affected by the development of energy policies and the siting of MEFs.

(iii) Given the fact that the siting of MEFs should be a long lead process, Member governments should make every possible effort to develop long-term LUP systems on either or both national and regional levels and to ensure that the siting of MEFs takes place within the framework of such LUP.

(iv) National energy research and development programmes should implicitly consider the siting implications of energy systems to be developed and to either identify additional areas of research or modify the relative emphasis given to new technologies under development.

Procedures

Regulatory procedures for the siting of MEFs are not simply administrative tools, but play an important role on the potential impacts of MEFs on the environment. They define the framework within which the political processes of impact assessment, interaction on interests, and final choice are taking place, as well as the quality and extent of participation of groups and individuals.

(i) Traditionally, the siting of MEFs was decided between the developer and the local authority. Such an arrangement can no longer provide solutions to the problems of siting.

(ii) Among the three levels of government which exist in most OECD countries, it is the regional government (authority) which is best placed to undertake the major responsibility in siting MEFs.

(iii) Regional governments should be given, or allowed to acquire, the necessary financial, human and other means to enable them to carry their responsibilities in the siting process.

(iv) Regional governments - in co-operation with national and local authorities - should develop long-term LUP programmes. Such programmes should be developed on the basis of, among other factors, general environmental assessments of the capacity of the region, or of areas within the region, to absorb impacts from industrial and other development.

(v) Sites for MEFs should be identified in the LUP programmes by co-operation among national, regional, local authorities and developers, and in accordance with long-term energy and economic growth projections. The possible environmental impacts from such probable developments should be identified on a national and regional basis.

(vi) Whenever a decision is taken that an additional MEF is needed, local planning authorities should ensure that:

(a) the sites proposed are in accordance with the indications of the long-term regional and local LUPs;

(b) that an environmental impact assessment is carried out.

(vii) The Environmental Impact Assessment will evaluate only those probable impacts which will be identified and agreed upon by the developer, the local and regional authorities and the local public and which have not been assessed previously.

When assessment is completed and the local public and interested groups have had an opportunity to review and comment on its findings, then it should be considered by the licensing authorities.

(viii) Given the fact that administrations require a large number of permits - water use, clean air standards, transport facilities, etc. - to provide a building and an operation licence, local authorities, through their land use planning services, undertake the responsibility to act as the co-ordinator in obtaining these permits, once the developer has put through an application for an MEF. This may simplify and enhance the siting process.

(ix) The "public utility" character of some of the MEFs should not give them special status when their siting is being considered given the fact that some of them are among the most polluting and/or most dangerous industrial installations.

Public participation

Public participation is not a panacea to wise selection of sites. It can, however, help in making the choice by reflecting the public's preference; i.e. by reducing the social adverse impacts. On the other hand the public (local, regional) may take a negative attitude to any development, if it is not given the opportunity to express its opinion and have it considered by the authorities.

(i) The public may have strong opinions - differing from those of its elected representatives - on two levels:

a) on a national level with respect to energy policies (e.g. the nuclear issue), energy R & D, etc.;

b) on the regional and/or local level on siting individual MEFs.

Although a complete separation of the two issues is not possible, it will be useful that Member governments structure their policy development and regulatory procedures in such a way as to allow their citizens to express their opinion on one issue with as little influence of the other as it may be possible.

(ii) Given the fact that the siting of MEFs:

(a) may affect groups of people as diverse as labour unions, environmental groups, industrial and business associations, etc.;

(b) may produce substantial socio-economic positive and negative impacts;

and recognising:

(c) that priorities and criteria with respect to environment and development may differ substantially among regional populations in a country;

(d) that particularly vociferous or strong interest groups may distort the issues;

Member governments may, whenever a decision cannot be reached otherwise and whenever the Constitution permits it, allow their regional authorities to organise referenda in order to obtain as clear and as unambiguous an expression of public opinion as is possible.

REFERENCES

(1) Energy Alternatives: A Comparative Analysis, Science and Public Policy Programme, University of Oklahoma, May 1975, U.S. Government Printing Office, No. 041-011-00025-4.

(2) An Analysis of the impact of refinery siting, Radian Corporation, prepared for the Environment Protection Agency, August 1974.

(3) Environmental Residuals Coefficients for the Project Independence Evaluation System Model, Hittman Associates, Inc., prepared for the Environment Protection Agency, July 1976.

(4) Scottish Development Department, Oil Terminals; Implications for Planning, Dec. 1975.

(5) World Energy Outlook, OECD, Paris 1977.

(6) Energy Alternatives: A Comparative Analysis, Science and Public Policy Programme, University of Oklahoma, May 1975.

(7) Ponder, W.H., Stern, R.D., McGlamery, G.G., "SO_2 Control Technologies - Commercial Availabilities and Economics". Third International Conference on Coal Gasification and Liquefaction, Pittsburgh, Pennsylvania, August 1976.

(8) Royal Commission on Environmental Pollution - Chairman Sir Brian Flowers. Sixth Report: Nuclear Power and the Environment, HMSO, London, 1976.

(9) Nuclear Regulatory Commission - Reactors Safety Study, Washington 1400, (1975), (The "Rasmussen Report").

(10) IAEA/NEA. "Siting Nuclear Facilities", Vienna, 1974.

(11) Nuclear Safety, 15,5 (September-October 1974), p. 519.

(12) Meier, P., "Sociopolitical Ramifications of Nuclear Energy Centres", BNL, 1977

OECD SALES AGENTS
DÉPOSITAIRES DES PUBLICATIONS DE L'OCDE

ARGENTINA – ARGENTINE
Carlos Hirsch S.R.L., Florida 165,
BUENOS-AIRES, Tel. 33-1787-2391 Y 30-7122

AUSTRALIA – AUSTRALIE
Australia & New Zealand Book Company Pty Ltd.,
23 Cross Street, (P.O.B. 459)
BROOKVALE NSW 2100 Tel. 938-2244

AUSTRIA – AUTRICHE
Gerold and Co., Graben 31, WIEN 1. Tel. 52.22.35

BELGIUM – BELGIQUE
LCLS
44 rue Otlet, B1070 BRUXELLES. Tel. 02-521 28 13

BRAZIL – BRÉSIL
Mestre Jou S.A., Rua Guaipá 518,
Caixa Postal 24090, 05089 SAO PAULO 10. Tel. 261-1920
Rua Senador Dantas 19 s/205-6, RIO DE JANEIRO GB.
Tel. 232-07. 32

CANADA
Renouf Publishing Company Limited,
2182 St. Catherine Street West,
MONTREAL, Quebec H3H 1M7 Tel. (514) 937-3519

DENMARK – DANEMARK
Munksgaards Boghandel,
Nørregade 6, 1165 KØBENHAVN K. Tel. (01) 12 85 70

FINLAND – FINLANDE
Akateeminen Kirjakauppa
Keskuskatu 1, 00100 HELSINKI 10. Tel. 625.901

FRANCE
Bureau des Publications de l'OCDE,
2 rue André-Pascal, 75775 PARIS CEDEX 16. Tel. (1) 524.81.67
Principal correspondant :
13602 AIX-EN-PROVENCE : Librairie de l'Université.
Tel. 26.18.08

GERMANY – ALLEMAGNE
Alexander Horn,
D - 6200 WIESBADEN, Spiegelgasse 9
Tel. (6121) 37-42-12

GREECE – GRÈCE
Librairie Kauffmann, 28 rue du Stade,
ATHÈNES 132. Tel. 322.21.60

HONG-KONG
Government Information Services,
Sales and Publications Office, Beaconsfield House, 1st floor,
Queen's Road, Central. Tel. H-233191

ICELAND – ISLANDE
Snaebjörn Jónsson and Co., h.f.,
Hafnarstraeti 4 and 9, P.O.B. 1131, REYKJAVIK.
Tel. 13133/14281/11936

INDIA – INDE
Oxford Book and Stationery Co.:
NEW DELHI, Scindia House. Tel. 45896
CALCUTTA, 17 Park Street. Tel. 240832

ITALY – ITALIE
Libreria Commissionaria Sansoni:
Via Lamarmora 45, 50121 FIRENZE. Tel. 579751
Via Bartolini 29, 20155 MILANO. Tel. 365083
Sub-depositari:
Editrice e Libreria Herder,
Piazza Montecitorio 120, 00 186 ROMA. Tel. 674628
Libreria Hoepli, Via Hoepli 5, 20121 MILANO. Tel. 865446
Libreria Lattes, Via Garibaldi 3, 10122 TORINO. Tel. 519274
La diffusione delle edizioni OCSE è inoltre assicurata dalle migliori librerie nelle città più importanti.

JAPAN – JAPON
OECD Publications and Information Center
Akasaka Park Building, 2-3-4 Akasaka, Minato-ku,
TOKYO 107. Tel. 586-2016

KOREA - CORÉE
Pan Korea Book Corporation,
P.O.Box n° 101 Kwangwhamun, SÉOUL. Tel. 72-7369

LEBANON – LIBAN
Documenta Scientifica/Redico,
Edison Building, Bliss Street, P.O.Box 5641, BEIRUT.
Tel. 354429–344425

MEXICO & CENTRAL AMERICA
Centro de Publicaciones de Organismos Internacionales S.A.,
Alfonso Herrera N° 72, 1er Piso,
Apdo. Postal 42-051, MEXICO 4 D.F.

THE NETHERLANDS – PAYS-BAS
Staatsuitgeverij
Chr. Plantijnstraat
'S-GRAVENHAGE. Tel. 070-814511
Voor bestellingen: Tel. 070-624551

NEW ZEALAND – NOUVELLE-ZÉLANDE
The Publications Manager,
Government Printing Office,
WELLINGTON: Mulgrave Street (Private Bag),
World Trade Centre, Cubacade, Cuba Street,
Rutherford House, Lambton Quay, Tel. 737-320
AUCKLAND: Rutland Street (P.O.Box 5344), Tel. 32.919
CHRISTCHURCH: 130 Oxford Tce (Private Bag), Tel. 50.331
HAMILTON: Barton Street (P.O.Box 857), Tel. 80.103
DUNEDIN: T & G Building, Princes Street (P.O.Box 1104),
Tel. 78.294

NORWAY – NORVÈGE
J.G. TANUM A/S,
P.O. Box 1177 Sentrum, OSLO 1

PAKISTAN
Mirza Book Agency, 65 Shahrah Quaid-E-Azam, LAHORE 3.
Tel. 66839

PORTUGAL
Livraria Portugal, Rua do Carmo 70-74,
1117 LISBOA CODEX.
Tel. 360582/3

SPAIN – ESPAGNE
Mundi-Prensa Libros, S.A.
Castelló 37, Apartado 1223, MADRID-1. Tel. 275.46.55
Libreria Bastinos, Pelayo, 52, BARCELONA 1. Tel. 222.06.00

SWEDEN – SUÈDE
AB CE Fritzes Kungl Hovbokhandel,
Box 16 356, S 103 27 STH, Regeringsgatan 12,
DS STOCKHOLM. Tel. 08/23 89 00

SWITZERLAND – SUISSE
Librairie Payot, 6 rue Grenus, 1211 GENÈVE 11. Tel. 022-31.89.50

TAIWAN – FORMOSE
National Book Company,
84-5 Sing Sung Rd., Sec. 3, TAIPEI 107. Tel. 321.0698

UNITED KINGDOM – ROYAUME-UNI
H.M. Stationery Office, P.O.B. 569,
LONDON SE1 9 NH. Tel. 01-928-6977, Ext. 410 or
49 High Holborn, LONDON WC1V 6 HB (personal callers)
Branches at: EDINBURGH, BIRMINGHAM, BRISTOL,
MANCHESTER, CARDIFF, BELFAST.

UNITED STATES OF AMERICA
OECD Publications and Information Center, Suite 1207,
1750 Pennsylvania Ave., N.W. WASHINGTON, D.C.20006.
Tel. (202)724-1857

VENEZUELA
Libreria del Este, Avda. F. Miranda 52, Edificio Galipán,
CARACAS 106. Tel. 32 23 01/33 26 04/33 24 73

YUGOSLAVIA – YOUGOSLAVIE
Jugoslovenska Knjiga, Terazije 27, P.O.B. 36, BEOGRAD.
Tel. 621-992

Les commandes provenant de pays où l'OCDE n'a pas encore désigné de dépositaire peuvent être adressées à :
OCDE, Bureau des Publications, 2 rue André-Pascal, 75775 PARIS CEDEX 16.
Orders and inquiries from countries where sales agents have not yet been appointed may be sent to:
OECD, Publications Office, 2 rue André-Pascal, 75775 PARIS CEDEX 16.

OECD PUBLICATIONS
2, rue André-Pascal, 75775 Paris Cedex 16
No. 40.863 1979
PRINTED IN FRANCE